Principles of Engineering Design

Vladimir Hubka

Translated and edited by
W. E. Eder
Associate Professor of Mechanical Engineering, Royal Military College of Canada, Kingston, Ontario, Canada.

Butterworth Scientific
London Boston Sydney Wellington Durban Toronto

All rights reserved. No part of this publication may be reproduced or transmitted in any form or by any means, including photocopying and recording without the written permission of the copyright holder, application for which should be addressed to the publishers. Such written permission must also be obtained before any part of this publication is stored in a retrieval system of any nature.

This book is sold subject to the Standard Conditions of Sale of Net Books and may not be resold in the UK below the net price given by the publishers in their current price list.

First published as *Allgemeines Vorgehensmodell des Konstruierens* by Fachpresse Goldach, Zurich, 1980

© Vladimir Hubka, 1980

First English edition published by Butterworth & Co (Publishers) Ltd, 1982

© Butterworth & Co (Publishers) Ltd, 1982

British Library Cataloguing in Publication Data

Hubka, Vladimir,
Principles of engineering design.
1. Engineering design
I. Title   II. Allgemeines Vorgehensmodell des Konstruierens.
English
620'.00425   TA174   80-42027
ISBN 0-408-01105-X

Filmset in 10/11pt Compugraphic English Times by
RDL., 26 Mulgrave Road, Sutton, Surrey.
Printed in England by Redwood Burn Ltd., Trowbridge

Author's Preface

Methodical procedures in design engineering, in spite of their relatively recent use, have taken on a wide variety of forms. Is this multiplicity only an external characteristic, or is it an integral part of the problem? This question represents a challenge, and has stimulated a search for the common denominator in the efforts to date.

One result of this search—the General Procedural Model of Design Engineering—can be viewed as an attempt to synthesise published opinions, contributions to discussions by students and designers, and experiences from teaching in colleges and in continuing education courses.

In preparing this work, attention was given to the following requirements:
- as far as possible, general applicability to the area of machine systems;
- as far as possible, consistent treatment of each step, and consistent terminology;
- a rational foundation for all steps and for its sequence;
- presentation of the fundamental knowledge that can assist understanding of the procedures;
- minimum of descriptive content, to ensure adequate clarity, and to preserve widest generality.

Based on these requirements, I hope that the model will assist both the student and the practising design engineer.

I owe a debt of gratitude for the development of this book to my friend, M. Myrup Andreasen.

Vladimir Hubka
Greifensee, April 1980

Translator's Preface

The present text is the result of two almost conflicting lines of thought. On the one hand, the method of clear logical progression with its tendency to suppress feelings and attitudes has determined the sequence and method of presentation of the basic ideas in this book. This has resulted in an attempt to prescribe a generally applicable abstract method, based on a generic description of circumstances and definitions. Only in later sections does this approach receive any modification to account for human differences, and for different fields of application. The work may therefore appear to be mechanistic, and excessively academic.

The second influence is the more liberal viewpoint of design engineers who put a premium on individuality, and who regard prescriptions of human behaviour, especially of such highly personal procedures as engineering design, as strait-jackets that can only inhibit creative endeavours.

Under these extremes of influence, the translator must stand with feet in opposing camps. It is essential at this stage of development, that both camps learn to understand one another, to find common ground, and to develop the strength of their profession. I have therefore undertaken to translate this book, but have attempted to make it more presentable to its English-speaking audience by adding explanations and examples within the text, as well as references to selected publications in English.

In preparing this translation, I have used some well-known general dictionaries, and a newly compiled dictionary of design terminology (the latter published as a companion volume to the original German text). I have also found it necessary to set up a further short dictionary, in order to maintain a fairly strict vocabulary.

In order to learn or modify a design approach, some prescription of sequence and method seems to be necessary, and systematic method should:
(a) provide active support for the innate abilities of the practising design engineer in his quest to synthesise intuition and knowledge,
(b) help to avoid that precipitous jump to the first available conclusion (which is seldom the best), and
(c) help to make the search for a solution to a technological problem more comprehensive, and the selection of the optimum more rational.

The main emphasis of this book is on mechanical engineering, the branch

Translator's Preface

that at present has the least developed methodological base in the English-speaking regions of the world. In continental Europe, much useful work has been done towards improving mechanical engineering design processes, one result is the present book. European work has received relatively little attention in the English-speaking world, just as English-language literature has been fairly poorly received in Europe—this seems to be a case of mutual blindness. I hope that this translation (with its additions) can open some eyes.

The general structure of this book is summarised in an addition to Chapter 1 and in Chapter 7. According to this rationale, the technical and material aspects receive earlier treatment to provide a discussion base for the following human activity. This procedure requires some discipline on the part of the reader, who should defer judgement on the text until a later chapter, when the difficulties of creative activity receive their due mention. It is particularly with this circumstance in mind that I have written various additional paragraphs, and have also chosen to translate phrases such as "is to be", a dictatorial prescription, into "should be", a strong recommendation. Nevertheless, I hope that I have been able to comply with the author's original intentions.

I wish to acknowledge the assistance received from the following persons:
Peter Hills and colleagues from Royal Military College of Science, Shrivenham,
Mostyn Kimber from Hatfield Polytechnic,
Tony Pennell from the University of Newcastle-upon-Tyne,
Geoffrey Pitts from The University, Southampton,
Noel Svensson from The University of New South Wales,
and of the publisher's editors, Don Goodsell and Peter Lafferty, in reading the English manuscript and making appropriate suggestions for its improvement. I have also received the active help of the author, Vladimir Hubka, who has seen most of my additions, and has expressed his wholehearted agreement with their content. The exchange of views has, I believe, benefitted both sides. His friend, M. Myrup Andreasen, has helped particularly with many of the terminology definitions. My thanks go to them, to others unnamed who have provided advice and support, and particularly to the staff of the Data Unit, Loughborough University of Technology, for their assistance, and for use of word-processing facilities.

W.E. Eder
Loughborough, June 1981

List of Tables

Table 1 Categories of Machine Elements 19
Table 2 Technical Processes: Classification System according to Levels of Abstraction 22
Table 3 Design Tactics: Survey of Methods 38-39
Table 4 Design Tactics: Survey of Working Principles for Design Engineers 40-41
Table 5 Welding Positioner: Assigned Problem Statement 78
Table 6 Welding Positioner: Design Specification (List of Requirements) 79-80

List of Figures

	Frontispiece	
Fig. 1	Definition and Representation of the Technical Process	9
Fig. 2	General Model of the Structure of the Technical Process	10
Fig. 3	Model of a Technical System	13
Fig. 4	Functional Structure of the Technical System	14
Fig. 5	Technical System: Survey of Classes of Properties and their Relationships	16
Fig. 6	Origination- and Life-Stages of Technical Systems	20
Fig. 7	Origination- and Life-Stages of Technical Systems	21
Fig. 8	Technical Systems: Classification System according to Levels of Abstraction	23
Fig. 9	General Model of the Design Process	31
Fig. 10	Structural Parts of the Design Process	33
Fig. 11	Application of the Degrees of Completeness and of Finality of the TS-Properties during Design Progress	35
Fig. 12	Survey and Contents of Design Documentation	42
Fig. 13	General Procedural Model of the Design Process	46
Fig. 14	The Design Process: Graphical Representation of the States of Technical Systems during Design Work	47
Fig. 15	Design Manuals (Catalogues): Survey and Examples	66-67
Fig. 16	Evaluation: Procedural Model of some Types of Evaluation	70
Fig. 17		81
Fig. 18	Welding Positioner: Technical Process (Block Schematic)	82
Fig. 19		83
Fig. 20	Welding Positioner: Functional Structure (Block Schematic)	84
Fig. 21	Welding Positioner: Functional Structure (Hierarchical Tree)	84
Fig. 22		85
Fig. 23	Welding Positioner: Morphological Matrix	86
Fig. 24	Welding Positioner: Concepts and Concept Sketches	86
Fig. 25	Welding Positioner: Preliminary Layout as Hand Sketch	88
Fig. 26	Welding Positioner: Dimensional Layout with Detail	88

Contents

Author's Preface
Translator's Preface
List of Tables
List of Figures

1	**General Introduction**	1
1.1	A Robinson Crusoe story	1
1.2	Structure of this book	2
2	**Technical Processes**	5
2.1	Human needs and their satisfaction	5
2.2	Basis of knowledge about technical processes	5
2.3	Trends in development of technical processes	6
2.4	Factors of processes	7
2.5	Secondary inputs and outputs of technical processes	7
2.6	Structure of technical processes	7
2.7	Operands used in technical process	8
2.8	General model of technical processes	8
2.9	Types of process	8
2.10	Generalisation and synopsis—statements TP	9
3	**Technical Systems**	12
3.1	Purpose and tasks of technical systems	12
3.2	Mode of action of technical systems	14
3.3	Properties and quality of technical systems	15
3.4	Secondary inputs and outputs, disturbances	16
3.5	Structure of technical systems, anatomy	17
3.6	Origination and operation phases of technical systems	20
3.7	Development of technical systems	21
3.8	Abstraction and classification of technical systems	22
3.9	Particular types of machine system. Branches of mechanical engineering	24
3.10	Synopsis of important statements about technical and machine systems—statements MS	24

4	**Design Processes**	27
4.1	Basis of knowledge of systematic design	27
4.2	Premise for validity of general design methodology	29
4.3	Tasks and methods of design research	30
4.4	General model of the design process	31
4.5	Structure of the design process	32
4.6	Design strategy, procedural model	34
4.7	Fitting the general model to particular conditions	36
4.8	Design tactics, methods and working principles	37
4.9	Representation during design	40
4.10	Working means in the design process	41
4.11	Synopsis of design process statements—statements DesP	42

5	**General Model of Methodical Procedure during Design—the General Procedural Model**	44
5.1	Requirements of the general model	44
5.2	Concepts of the model	45
5.3	Explanations of the model	48
5.4	Particular steps in the model	48
	Stage 1—Elaborate or clarify the assigned problem	49
	Stage 2—Establish the functional structure	50
	Stage 3—Establish concepts	53
	Stage 4—Establish preliminary layout	57
	Stage 5—Establish dimensional layout	59
	Stage 6—Detailing, elaboration	61
5.5	Generalized basic operations in the design process	62
	5.5.1 Elaborating the assigned problem	63
	5.5.2 Searching for solutions	65
	5.5.3 Evaluating and deciding	68
	5.5.4 Providing and preparing information	72
	5.5.5 Verifying/checking	74
	5.5.6 Representing	76

6	**Case Study: A Welding Positioner**	77
6.1	Elaboration of assigned specification	77
6.2	Establishing the functional structure	81
6.3	Establishing the concept	85
6.4	Establishing the preliminary layout	87
6.5	Establishing the dimensional layout	87

7	**Summary**	89

8	**Bibliography**	90

9	**Glossary**	**95**
9.1	Narrative definitions	95
9.2	Key-word definitions	98

Index 113

Chapter 1

General Introduction

1.1 A ROBINSON CRUSOE STORY

Let us observe the behaviour of a survivor from a shipwreck who has to exist on a deserted tropical island. If he is hungry, and there is a banana tree before him with fruit that is easily reached, the circumstances offer no technical problems. Problems only emerge if the fruit is too high up, or if the tree stands on the other bank of a river. Then an obstacle exists between him and his objective. How does Robinson Crusoe behave in this new situation? If his feeling of hunger is not too great, or if he can find other means of nourishment, or if he is afraid to collect the bananas from the tall tree, he could deny himself the pleasure of the fruit.

If on the other hand he has set himself the target of obtaining those bananas—influenced (motivated) by certain well defined conditions—there are two possibilities: either he tries arbitrarily (without plan) to obtain the bananas by any available method or means, for instance he may throw stones, take a stick, or even try to shake the tree; or, on the other hand, he may think and reflect. He investigates the situation, and establishes which are the obstacles and possibly their causes, and also what resources are available to him. Then he quantifies the conditions, e.g. the distance to the bananas, or the depth of the river. He examines whether he or someone else may have been in a similar situation, and by which means and methods they reached their target. He tries thereby to utilise existing knowledge and experience. Some of the possibilities can be rejected immediately, for instance, if a potential tool or method is not available to him. Other more appropriate means are scrutinised, and he selects the one that appears best suited. He then realises (plans and makes) any new tools, and implements (uses) the adopted solution, and can then either pleasantly satisfy his hunger, or an unsuccessful attempt forces him to renewed thought based on an analysis of his experience. Reaching the target is usually not the end of the endeavour, rather he will continue to think about the problem to determine whether the target can in future be reached in a simpler and quicker fashion, with less use of resources and energy.

This case, in which Robinson Crusoe has thought out, realised and implemented a plan of action using certain available means, is relatively simple and easily understood. The situation becomes more complicated if the hungry Crusoe is sitting on the sea shore and considers how he can satisfy his hunger. Sticks and stones are not the only possible means for solving his problems. He considers—having decided to relieve his hunger by trying to catch a fish—that some means of keeping afloat above a likely location for fish (e.g. a boat) would be needed. Such a device is not as easy to find in nature as a stick! Robinson Crusoe must therefore develop a concept of a boat that can be transformed into reality using the limited means at his disposal. He designs, and in the process he may sketch in the sand. Then he builds the boat, thereby realising his concept, and finally he launches it to implement his chosen solution to the self-assigned problem. During manufacture various other problems arise as consequences of the needs of production processes. He requires, for instance, an axe (a means of working wood), but only finds various pieces of ironwork that came from a ship. He must transform these workpieces by some appropriate process into a desired form, which means that he must search, based on his knowledge of metal forging technology, for physical effects (and methods of realising them) such as heating, exerting pressure, and cooling. In realising these partial effects (which in sum produce the total effect of transforming the metal into a different and better form for his job-in-hand), he encounters further problems to be solved.

After completing his boat, a typical situation arises. Boat, axes, forging hearth, and hammers, everything that he has manufactured, are not only usable to obtain the pre-planned effects, but can be employed for a much wider range of tasks. The axe, for instance, can be used for the processes of hunting and defence, the boat for leaving the island, or as a sleeping area, or a roof (by turning it upside down on dry land). Crusoe has thereby increased his resources of technical means, and now has various additional physical effects available to him.

1.2 STRUCTURE OF THIS BOOK

The above example is intended to show the importance of design as a general activity for human progress, and particularly for the satisfaction of recognised needs. This process was historically based on experience and craft abilities. Such an approach tends to break down once the range of collected and abstracted knowledge, and of available devices, becomes sufficiently large. The knowledge immediately available to an individual is very limited, even if he (that chauvanistic expression, but we intend no harm, *she* can do even better at times!) has a number of books on his shelves (assuming, of course, that language and writing were sufficiently developed).

In an age of high and rapidly increasing technology, it seems no longer feasible to rely only on experience from which to learn how to design. A separate literature has grown around this subject, in an attempt to improve the human being's mental processes, and to make him more effective at solving his technical problems (e.g. 45, 46, 52, 62). This book is one such attempt.

Based on various considerations, by theorising and abstracting, the author has tried to create a clear distinction between different aspects of a technical device. His background is largely in mechanical engineering, but the resulting knowledge and procedures can be readily transferred (by analogies) to other fields of endeavour. In developing his ideas, the author has found it necessary to formulate a limited vocabulary of terms, and to apply it in a strict and consistent way. Parts of this vocabulary are now well established as a recognised design terminology in the German-language literature. A set of equivalent English terms has been developed by co-operation between a number of authors, but has not yet reached the state of certainty that surrounds the original language of this book.

With the present book in its translations we aim to introduce this vocabulary and terminology simultaneously into a number of languages. A companion volume published in six languages contains a set of formal definitions. A taste of the formalism of this work has crept into the previous section, further condensed explanation may be found in two forms in Chapter 9.

Design is a very personal activity, and can probably only be performed by one person as an internal and somewhat subjective process. '.... because, let's be honest, even the engineer is allowed to have a heart, rarely, to be sure, and then naturally only in his spare time.' (Jaro Zeman, ÖIZ, 23 Jg., H. 5, p. 176.) Most projects are too large. One person cannot hope to complete a typical project in a reasonable time, therefore that design activity must be made (a) as rational and effective as possible for the one person, but also (b) as open as possible to inspection for other co-operating persons (teamwork). The results of the design activity must be put down on paper (or some other permanent recording medium), so that the design engineer, and all associated persons, can see, review, criticise, evaluate, decide, and do all the activities needed to create and make a new project.

One way of improving the designer's effectiveness is to explain the basic physical and mental operations that he uses, especially by relating this knowledge to verifiable psychological evidence. A second way is to lay down a sequence of steps and stages that can be followed in an ideal case (and therefore modified if the situation happens not to be ideal), and to recommend the documentation that will keep the record of past activities, and provide the basis for the future tasks. This book sets out to provide both the sequence of steps and the explanation, with the purpose of describing them, but also (with knowledge of human limitations) of prescribing. It is in a very condensed form, but includes references to other literature. Many of the ideas put forward are probably unfamiliar in this form in the English-speaking part of this world.

The procedures and methods presented in this book are intended to give active support to the design engineer, not only to his knowledge, but also to his intuition. If he is to avoid jumping to the first conclusion that presents itself, he should be prepared to survey the problem and its implications with utmost care, and then to carry out a comprehensive search for likely solutions, before using his well founded judgement to select the best. The highest probability of success is likely to come from a methodical approach, added to design flair.

This book is structured to lead the reader through a number of related concepts. Chapter 2 defines the nature and scope of technical processes

(TP). Chapter 3 deals with the more tangible aspects, technical and machine systems (TS and MS). Chapter 4 treats the human design engineer as an analogue of a technical system, and the design process as an analogue of a technical process. It covers the design stages, methods and procedures, and documentation. Chapter 5 presents a complete model of the design process, from first problem assignment in a brief statement, to a completely documented solution ready to be made. The design of a welding positioner is given in Chapter 6 as a case study, related to the methods and procedures of the earlier chapters. The book ends with a short summary, a bibliography consisting of works of German-language origin, augmented by English-language literature, and a glossary of terms and their definitions as used in this book.

Chapter 2

Technical Processes

2.1 HUMAN NEEDS AND THEIR SATISFACTION

Human society has many *needs* (requirements or desires, real or imagined), and is constantly developing new ones (either through becoming aware of them, or by being persuaded that they exist), as is clearly shown by the Robinson Crusoe example. Such needs have differing *character and significance,* when compared to the base values of the absolute necessities required for sustaining life. In this context a discussion of the relative nature of 'absolutely essential needs' is undesirable, we will confine ourselves to an investigation of the changes, and their causes, relevant to the satisfaction of needs by technical processes.

Such ethical questions as whether 'satisfaction' is to be given to all human beings, or just to a few are also beyond the scope of this book, as are moral, political or conscientious objections levelled at certain products by interested persons of one or other conviction, (e.g. armaments, herbicides, pesticides, alcohol, drugs, welfare facilities, council housing, etc.). Such products still 'satisfy' the human needs of a part of the population, and whatever the political implications, the process of designing and creating them remains a common feature.

2.2 BASIS OF KNOWLEDGE ABOUT TECHNICAL PROCESSES

Man observed nature, and the changes in his material environment that occurred as a result of natural processes, long before he began to organise artificial processes. His technology was founded on known and substantiated *natural laws*. In other words, with the help of existing knowledge of observed physical, chemical and biological phenomena (effects) and the rationalised, and codified experiences (laws) abstracted from those observations he was able to realise certain changes.

In this context it was not really necessary for him to be able to describe the phenomena by means of words, and especially not to give any experimentally or theoretically validated explanations for the causes of the observed phenomena. Experience allowed him merely to associate cause and effect (not always correctly), and later to rationalise this experience, even if this induced him to use somewhat irrational rituals, (e.g. to appease an unknown higher power). Qualitative prediction by the human being showed understanding: quantitative prediction followed at a much later historic stage, and mathematical models and procedures came much later still.

If Robinson Crusoe needed a tree for his boat, then he remembered that trees fell during a forest fire, or were felled by beaver teeth. The human being can use these experiences, and other perceptions and observations. Most changes in nature occur over long periods of time, assisted by rain, frost, etc.; but the work of nature is usually too slow for humans. It is not much use waiting for nature to fell a tree, partly because the material deteriorates with time, and partly because nature's timing is unlikely to be convenient. The artificial transformation processes that man desires (e.g. from a growing tree to the raw material for his boat) should occur in a sufficiently short time period. The human tends to replace slow, uncontrolled natural processes by his own more effective planned and orderly artifices.

2.3 TRENDS IN DEVELOPMENT OF TECHNICAL PROCESSES

Development of technology is accompanied by development of a vast variety of processes and technical means. The word 'technology' is here intended to cover not only the science of practical or industrial arts (Concise Oxford Dictionary), but also the experimental (craft-based) know-how that in part defies description by scientific means. This statement is therefore intended to convey all the functions of any device, and all the functions of the means for manufacturing that device, including the functions of commerce, transportation, and management needed to realise and distribute it. This interpretation is closer to the German word 'Technik' than the dictionary definition of technology in the English language. In a narrower sense, the word 'technology' can also be interpreted merely as the means and knowledge needed to perform manufacturing operations.

The human being is progressively replaced in various functions as a result of technological developments: he uses a variety of *tools;* his force capability and power output are limited so he is replaced by *mechanical prime movers* (first technical-scientific revolution); but he is also slow and unreliable: the *control* function is taken over by machines (second technical-scientific revolution).

The direction of developments indicated here, namely *instrumentation,* mechanisation (energy application), and *automation* (control application) are not only historically important and noteworthy; they also characterise certain stages during engineering design in which one makes decisions about the distribution between man and machine of the propulsion (drive) and control functions for the technical means to be designed.

2.4 FACTORS OF PROCESSES

A technological development as described in the previous section is inconceivable on Robinson Crusoe's island; but if we return to that island's conditions, further factors in process realisation can be investigated. Questions must be posed with respect to *space* (environment and location), and *time* (epoch or period). The assumption of an island in a tropical region provides more favourable boundary conditions to Crusoe's problems than assuming an island in the northern oceans (where for instance fruit may not grow). The *knowledge* and *organising talent* possessed by our Robinson Crusoe are further factors that influence his decisions about the processes that are realisable for him, and their operating parameters. Robinson Crusoe, a few hundred years ago, had a different spectrum of knowledge compared to a modern castaway.

2.5 SECONDARY INPUTS AND OUTPUTS OF TECHNICAL PROCESSES

Further undesirable *inputs* to the technical process can come from the environment, and these act mainly as disturbances, e.g. rain on the fire used by Robinson Crusoe to heat the metal. The process itself can also produce undesired *outputs:* the fire can spread to the forest; the bananas (or their consumer) leave their peel behind; manufacture of the boat produces wood shavings. The ecology is usually adversely affected by these output effects of the process, they are influences on the environment.

2.6 STRUCTURE OF TECHNICAL PROCESSES

Experience shows that every process can be sub-divided into *smaller component processes,* e.g. operations or hand movements. Building the boat consists of a large number of operations, connected with the preparation of raw materials and of tools, as well as with manufacture, testing or inspection.

Each process has a composite structure made up from its *partial processes and operations.* Three major phases may be recognised from the time-sequence (progress) of the work flow, namely *preparation, execution* (performing the transformation), and *conclusion* (review, recording, reporting, etc.). A different classification results from the type of actions needed to perform the process. A purposeful *main* (or work) transformation is accompanied by various auxiliary, propulsion (drive, energy conversion), and *control* processes, and in some cases also *connection* and *support* operations, and these are combined with the primary transformations to form a total process. If the human being himself performs the process, or the work of realising the process, he is usually unaware of these conceptual divisions.

2.7 OPERANDS USED IN TECHNICAL PROCESSES

Operands are subjected to transformations in the technical process, and appear as its inputs, throughputs and outputs. Three or four classes of operands can be recognised: *material* and *biological objects, energy,* and *information.* These classes always occur in combination; they cannot be isolated, but one of them will usually dominate the others for any applications or considerations. The categories of transformations show four basic varieties: changes of *structure* (internal changes, e.g. crystal lattice), of *form* (external changes, e.g. shape, or a liquid-to-gas transition) of *time,* and of *location* (position in space).

2.8 GENERAL MODEL OF TECHNICAL PROCESSES

The ideas concerning technical processes (TP) may be integrated into a general model, Fig. 1. This figure—as with every model of this kind—is intended to give a condensed, clear survey of the subject: in this case about Technical Processes (TP). The model also permits an overview of the *complete system* and its components, particularly:

(1) The operand (e.g. workpiece, object undergoing transformation within the TP)—horizontal flow through the rectangular symbol in Fig. 1
(2) The system of human and technical means that deliver the required physical effects to the TP (see Chapter 3)—oval symbol
(3) The processing component: operations that perform the transformations within the working process (technology)—rectangular symbol
(4) The process factors—effects acting on the TP that do not originate from either the human or the technical systems
(5) The secondary inputs and outputs

2.9 TYPES OF PROCESS

With respect to the different classes of operands, one can speak of material-, energy-, and information-processing operations. For those processes in which the human being appears as operand (e.g. education and training as experienced by the learner, or the human being as a passenger in a transportation process) a collective term is not available. For other operands, four basic types of process may be characterised by the categories of change mentioned above, i.e.:

Processing Operations: processes that primarily change the structural (internal) properties of the operands.
Manufacturing Operations: processes that primarily change the form (geometry, shape, constitution) of the operand.
Transport: processes that change the location of the operand in both space and time.
Storage: processes that only change the time location of the operand.

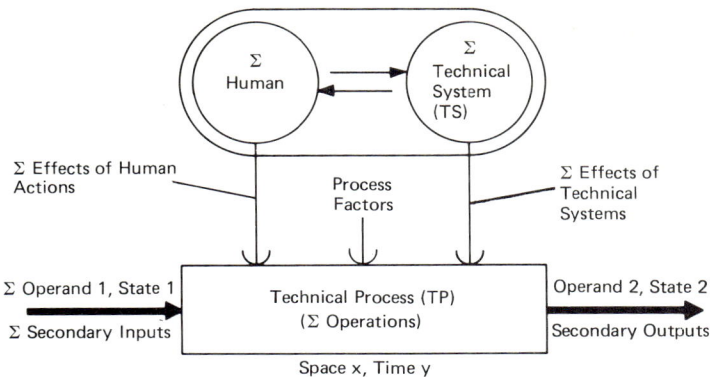
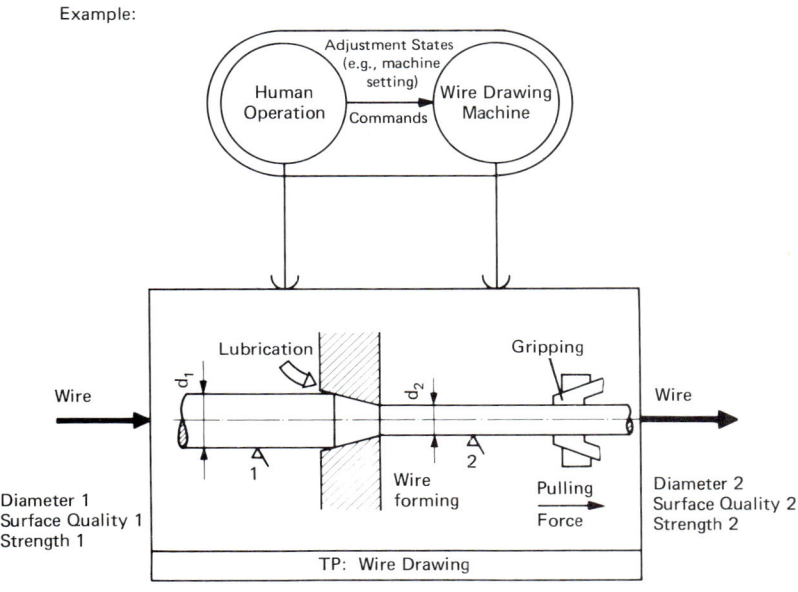

Fig. 1 Definition and Representation of the Technical Process.

These terms are familiar when applied to materials, but may also be extended to include energy and information processing operations, to make the terminology more precise.

2.10 GENERALISATIONS AND SYNOPSIS

A hierarchy of abstractions for technical processes may be developed by making individual elements (the operands, the technology, factors, etc.) of the system of processes progressively more concrete. The following statements summarise the main concepts about technical processes as developed in this chapter:

TP 1: A *technical process* is an artificial process in which the states of material and biological objects, energy, and information are changed (transformed) in a planned fashion, under the influence of the *effects received from human beings and technical means* (technical systems) (Fig. 2).

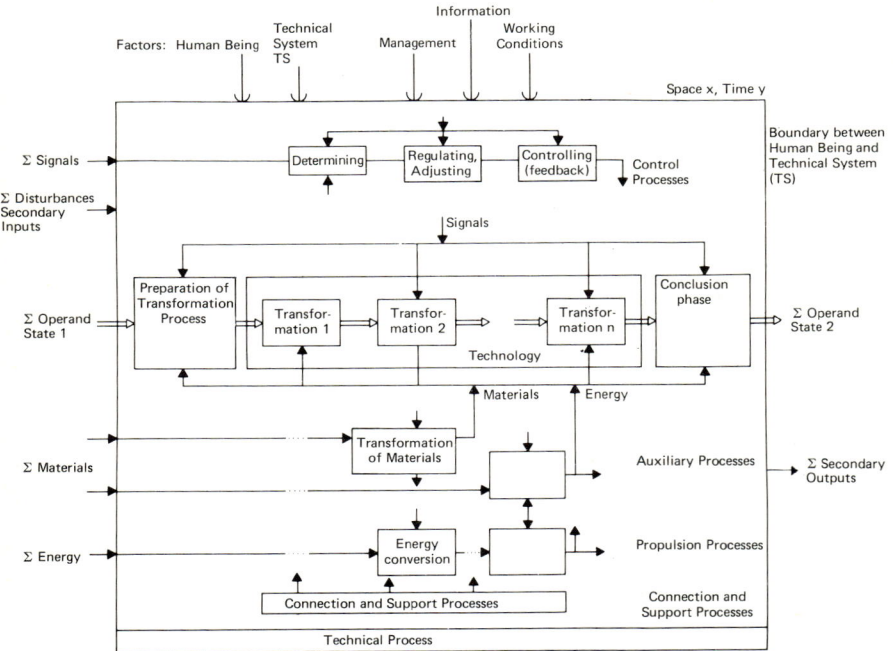

Fig. 2 General Model of the Structure of the Technical Process.

TP 2: The *states* (sum of properties) *attained* by the operands (objects subjected to change, i.e. materials, energy and information) as outputs of the processes serve to *satisfy human needs* of all kinds, either directly (as consumer goods, foods, etc.) or indirectly as technical means (work devices) that deliver effects necessary for performing technical processes.

TP 3: The *change of state* of the operand may occur in various ways. Depending on which of the *Technological Principles* (taken from *knowledge of physics, chemistry or biology*) are selected, one may establish the *technology* (in its narrower sense) as the sum of all the necessary operations, and their sequence. This in turn enables one to establish the structure of the Technical Process (TP) and the type of effects that must be exerted on the operands; this then permits one to establish the requirements placed on the technical means.

TP 4: The *structure of the TP* consists of partial processes and operations (down to the detail of movements). Two different classifications are used:
- a process (partial process, operation, used as a generic term) generally contains a preparation, an execution, and a conclusion phase.
- the main processes (conditioned by the purpose of the TP) are accompanied by auxiliary, propulsion, control, but also by connection and support processes.

TP 5: As stated in definition TP 1, certain of the effects applied to the operands are exerted by human beings, others by technical means. According to the *distribution of effects (actions) between the human beings and the technical systems* (*means*) one can distinguish:

(a) Manual processes (craft operations) with a preponderance of human action for both energy and control effects.
(b) Mechanised processes in which the technical system takes over the effects of supplying energy.
(c) Automated processes in which the technical system takes over most of the control effects.

TP 6: The sequence of events of the process, its output and economics are influenced not only by the effects of human actions and of the technical system, but also by other factors. These include the current level of knowledge (information, state of the art), the method of control of the process (particularly with respect to its organisation, its management and the form of its work procedures), as well as the environment conditions (in the narrower sense of space and time conditions) under which the process is required to proceed.

TP 7: Beside the desired and planned inputs and outputs, there exist *undesired* inputs (e.g. disturbances of the TP) and outputs (e.g. waste, disturbances of the environment).

TP 8: The *model of technical processes* that reflects all the points discussed in this chapter is shown in Fig. 1. This formal representation offers an overview, orientation and systematisation of work, and thereby can help to avoid omissions of some aspects. This model can be interpreted with the help of statements TP 1 to TP 7.

TP 9: *Systematics of TP*. Technical Processes may be classified according to various viewpoints:

(a) according to class of operand: processes that transform material, energy, or information
(b) according to the category of change: processes that transform the operands with respect to structure, form, location (in space), or time
(c) according to the hierarchy of abstraction (see Table 2): process phylum, class, family, or genus (basic process), specialised processes.

Chapter 3

Technical Systems

Most people assume that the only items that represent technology are technical systems (i.e. mechanisms, such as machines, plant, equipment, apparatus, cars, etc., as mentioned in Section 1.2). They overlook the role of technical processes with their transformation duties.

3.1 PURPOSE AND TASKS OF TECHNICAL SYSTEMS

In general, a technical system delivers some *effects* to the process (as well as those delivered by the human being) that contribute to the transformation (change of structure, form, location in time or space) of the operands (the materials, energy, or information throughput) in a technical process. One cannot strictly say that a universal centre lathe (a technical system) has the purpose of imparting form to a rotationally symmetric workpiece. It is only capable of holding and rotating the workpiece, and of holding the cutting tool and moving it within a plane (i.e. the combined effects of certain portions of a lathe). All these movements occur only as a result of commands and control inputs by the human operator, which are entered by him either directly into the lathe, as a manual craft operation, or indirectly through a computer controller to deliver the required sequence of movements and to check that the desired lathe operations were performed.

If those capabilities of movement of the lathe are used, it is possible to wind a helical spring on that lathe (i.e. a different technical process, a different transformation). The effects delivered by the technical system take place at an action locality (space, surface, or line), in the case of the lathe at the rotating centre (the spindle nose) and at the support slideways which guide the tool movement, as effectors of the technical system (TS) 'lathe'.

The purpose of a technical system is to transform certain well defined *input quantities,* particularly material (e.g. auxiliary materials), energy, and information (e.g. commands), into *desired effects* (output quantities) in space and time (e.g. position, movement, velocity, force).

We are consequently dealing with *two separate processes* of transformation that should be distinguished in spite of apparent similarities. In one case, materials, energy and information (the operands) are transformed *within the TP* over a period of time from an initial state towards a desired state (Fig. 1, the flow through the rectangular box). This transformation is based on a selected technology (the technological principle), and is performed with the assistance of and under the effects exerted by human beings and technical systems. In the other case, the inputs to a *technical system*—materials, energy and information—are transformed at a single point in time into effects or actions (outputs of that TS) of the desired type (e.g. movement, force, heat, cooling, protection), at the desired time and place (i.e. at the action localities, the downward connections between the oval and rectangular symbols in Fig. 1). This latter transformation occurs according to the selected mode of action of the technical system (following the action principle, as shown in Fig. 3). A similar view is apparent with respect to human designers in Figs. 2-2 and 3-1 by Krick (67).

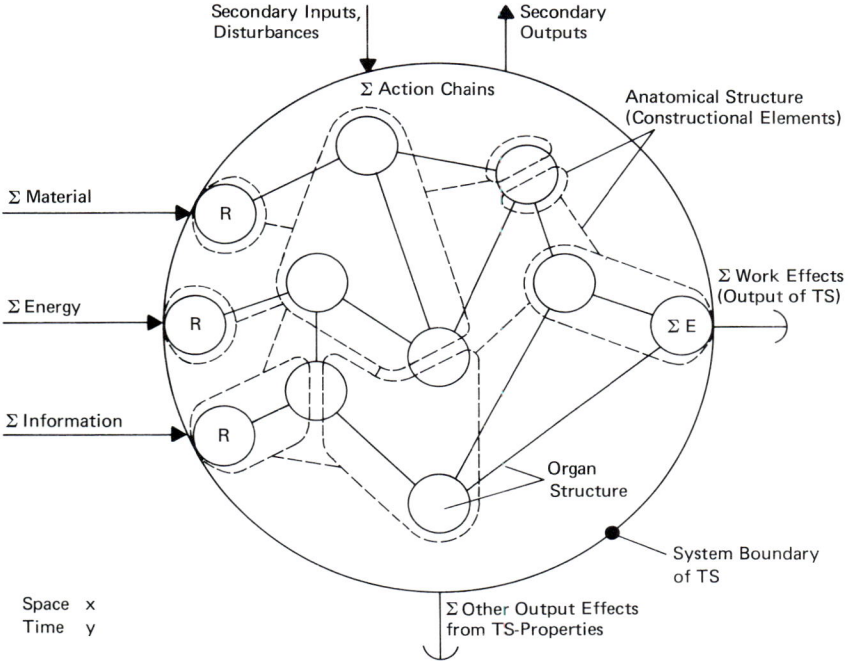

Fig. 3 Model of a Technical System.
  R: Receptor. E: Effector

A different *possibility of modelling* exists, in which the operands are considered in a combined input-technical-system and the necessary transformations take place inside that machine, i.e. the two processes considered separately in the last paragraph are now combined into one. This interpretation permits us to represent power machinery and automatics in a realistic and easily perceived way, but carries with it a danger that the two types of process are not clearly distinguished: (a) transformation of operands, and (b) mode of action of the technical system. It is probably sensible to use this form of realistic model at a more concrete level of the design process, and in a transition to a particular class of technical systems.

According to statement TP 4, the *transformation effects* (main output actions of the TS) are accompanied by obligatory *auxiliary, propulsion, control* and also *connection* and *support effects*. Each transformation demands preparation, execution, and conclusion actions. These insights permits one to assemble a general model of the *functional structure,* as a collection and arrangement of the obligatory output effects (functions) of a technical system to be designed (Fig. 4). Clearly the functional structure must be established at various levels of completeness and concretisation, depending on the stage of progress of the design work. This should be done preferably before the technical system, and its physical parts (anatomical structure and constructional elements), have been established at that stage of progress.

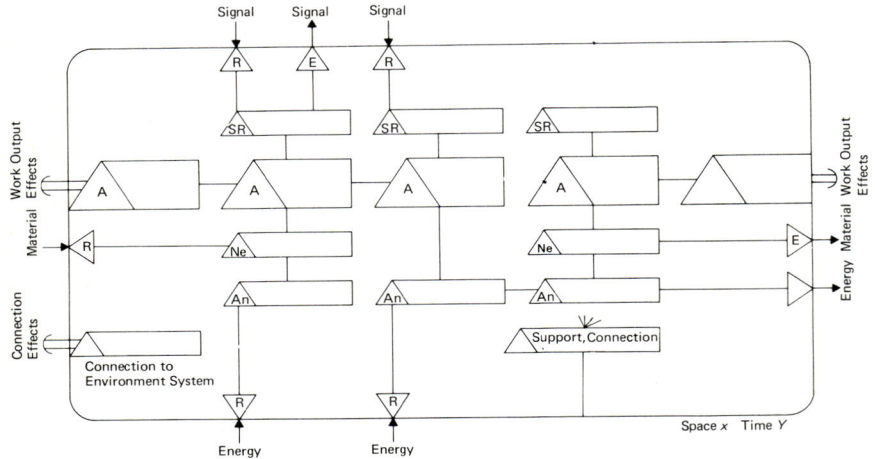

Fig. 4 Functional Structure of the Technical System.
 A: Work Function, Ne: Secondary Function, An: Propulsion Function, SR: Control (Regulating or Feedback) Function, R: Receptor Function, E: Effector Function
 NOTE: Symbols as shown on Frontispiece

3.2 MODE OF ACTION OF TECHNICAL SYSTEMS

The capability of technical systems to produce certain well-defined effects (outputs) from their inputs (materials, energy, information) is determined by a definite anatomical structure. This latter is an ordered collection of structural and manufacturable elements that have an active relationship with one another. It is this active relationship (the *mode of action,* or functional connectivity) that determines in which way and by what means the output effects occur, and that is commonly termed 'functioning' of the machine, i.e. an answer to the question: how does this mechanism operate? In consequence, the input-output relationships of a TS is termed its technical function (5). The mode of action of a whole technical system rests on that of an action chain (or chain of causality): every cause, starting from an input, results in a certain effect (action). The technical system is usually regarded as deterministic, even though all processes are known to be random (statistical, stochastic) in their detail. Usually, these variations are sufficiently small to be ignored, but occasionally a transfer and accumulation of small individual errors can result in large overall variability, and can make the concept of causality invalid.

When designing the desired action chain, the design engineer uses his knowledge of phenomena from various areas of the natural sciences, e.g. mechanical, electrical, chemical, or biological. The statement that 'a hoist works on a hydraulic principle' indicates from which area of knowledge and science the mode of action has been derived. Generally one can find a number of different classes of action modes within each action principle. We have here encountered a noteworthy general design-related characteristic of technical systems.

3.3 PROPERTIES AND QUALITY OF TECHNICAL SYSTEMS

The ability to exert output effects is not the only significant property of technical systems. They must above all perform these effects with the necessary operating parameters (e.g. power, speed, travel distance) within the envisaged environment, with ergonomically acceptable facilities for human operation, satisfactory appearance, transportability, and other properties or groups of properties. There are many of these properties (before Reuleaux they were thought to be of infinite number). Analysis and experience show that each technical system *possesses* (is the carrier of) a *range of similar groups of properties,* but in differing states of embodiment and value. The properties exist, even if they have not been consciously incorporated.

It has been shown (8) that all observable properties of a technical object depend on a class of basic properties, termed *design properties.* Therefore, not only the ability to produce effects, but also other properties such as strength, life, safety, appearance, transportability, etc. are caused by the *constructional elements* and their design characteristics. All these properties depend in particular on the form, size, material, manufacturing methods, surface properties, tolerances, arrangement, and state of assembly, etc., of these elements. This basic regularity permits us to consider the process of designing as a search for appropriate design properties (Fig. 5).

The various properties of a specific technical system appear in certain definite *states of embodiment,* with appropriate qualitative or quantitative measures (magnitudes, values). Considering colour (a general characteristic), an object may be red, blue, white, etc. (the state of embodiment of that characteristic); it is X cm high (a property), measured according to an agreed scale. Whether a particular state of embodiment, or the value of a property, is beneficial or acceptable in a specific case can usually be determined within close limits, most of the desired values are specified in the design requirements. Comparing the values associated with the

16   Technical Systems

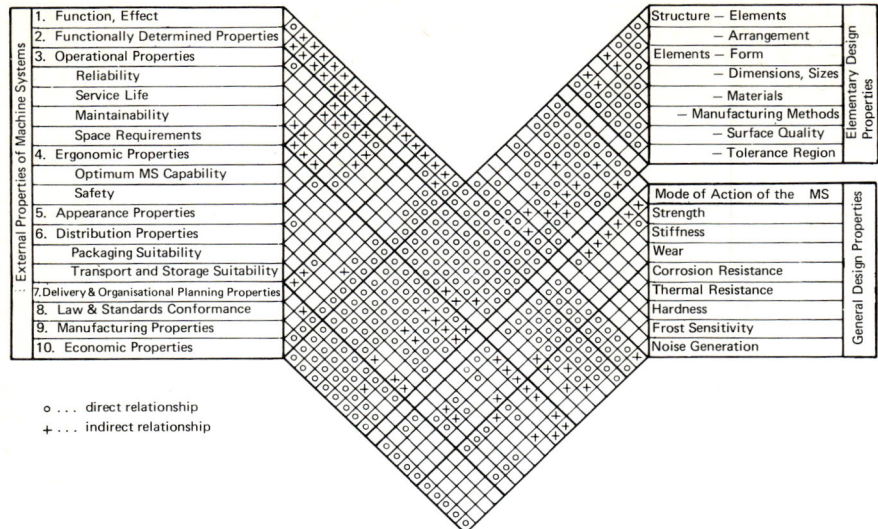

Fig. 5 Technical System: Survey of Classes of Properties and their Relationships.

individual properties of an existing or designed system with those of a previously defined value set (e.g. those given by the design requirements) permits us to make a judgement of the quality of a system, i.e. to perform an evaluation.

According to the number and grouping of properties that are used as criteria in performing such an *evaluation,* one obtains either an individual value (based on one property), or a single composite value which may represent the technical, economic, or total value. The maximum attainable quality of the object can be characterised either by the ideal, or by the desired value.

Combining a number of different values of properties into a total value is not without problems, and all known methods have disadvantages. Some of these are discussed in Section 5.5.3, and in the literature (55). It must be emphasised that combining different measures into a single criterion value tends to obscure the interactions of causes, and create a simplistic environment for the human decision process. The result is frequently a decision with only short-term benefits that can overlook the long-term necessities. Decisions should, where possible, be taken on the basis of a larger number of criteria, each with a survey of sensitivity to its influencing parameters, but this procedure can lead to a loss of objectivity. In addition, many of the criteria are inherently subjective (or have a substantial subjective content), and whilst quantification is desirable, it may not be easy to achieve the necessary degree of reliability or consistency (e.g. over a period of time, and between different assessors).

3.4 SECONDARY INPUTS AND OUTPUTS, DISTURBANCES

Technical Systems do not work in an isolated space, nor only over a definite period of time, neither are they protected from the influences of the surroundings, e.g. moisture, vibration, heat or light. A TS is in its turn always an element of a larger system. During the design process one should create a *critical model of the environment* and thereby wherever possible model all the likely influences on the technical system. The properties of the system should then be chosen to take account of the environmental influences (effects of the environment on the system), i.e. to resist the disturbances, and to avoid adverse influence on the function and other desired properties of the TS. The behaviour of the system must be ensured, even when disturbances occur.

The technical system (TS) that produces the desired output effects can only in rare cases be realised to an ideal state, when environmental conditions are considered. Noise and vibrations are also generated, and waste materials are produced. These secondary outputs, dependent as they are on the inputs to the technical system, its modes of action, and other factors, must also be investigated. The aim may be to avoid such emissions, to reduce them to an acceptable minimum (if necessary by additional functions), or to use them (e.g. by recycling, inside or outside the system).

3.5 STRUCTURE OF TECHNICAL SYSTEMS, ANATOMY

Technical systems contain as their structural components a number of manufactured or anatomical elements of varying complication. If we restrict the discussion to machine systems, the anatomical structure is composed of different machine parts or physical assemblies (groups of parts), that may be classified according to various aspects, e.g. design features, form, or other characteristics.

From a different viewpoint, that of function, various groupings occur that are distinguished by performing certain functions (producing certain effects) between or within components. In analogy to biological systems, the term 'organ' may be used for such a functional unit, or function-carrier. The essential features of each organ are those spaces, surfaces or lines that are the localities where the necessary effects take place, with little regard for the materials or objects needed to provide physical support to these *action localities*.

The *organs of a technical system* usually interact in an extremely complicated and interwoven way, and particularly between the boundaries of physical parts. It is usually advantageous for the subsequent process of establishing (imagining, thinking out, visualising and determining) an anatomical structure to use meaningful groups of parts, with clear boundaries in space, as assemblies. Some of these boundaries will constitute parts of the action localities contained in a particular organ, others will not be restrained by functions of the TS, e.g. manufactured surfaces of components that are only in contact with a non-functional environment. In practical terms, these assemblies form the subject of the group assembly drawings.

If, for instance, the cross-slide of a lathe forms an assembly (a sub-assembly for the whole lathe) then the function 'cross guidance' and the organ that performs this function are only partly represented in this sub-assembly, namely as the surfaces of the bearing pads. The mating surface, the guideway, is situated on the carriage, which is itself an assembly of physical parts containing various other (not necessarily complete) organs.

Technical systems usually exhibit a hierarchical arrangement. We use the term *'degree of complication'* of a system to define its position within a class level of the hierarchy. If we consider that machine systems can be arranged in four basic degrees of complication—plant or equipment, machines, assemblies, and parts—then the anatomical structure of a TS can always be sub-divided into machine systems (MS) of lower hierarchical order. In each of these hierarchical levels further sub-divisions can be generated (e.g. assemblies, and sub-assemblies).

A number of the components of technical systems recur in identical or similar construction in many existing machine systems, these are known as *machine elements* (ME). They may be single parts (1st level of complication), or functionally composite components consisting of two or more parts (2nd level). Classification of machine elements in the literature is variable: in technical books one can find consistency only at a basic level (screw, bearing, shaft, lever, wheel, etc.); usually the enumeration is incomplete. An orderly arrangement according to functional purpose and action principle is shown in Table 1.

The various *organs* (functional groups, or function-carriers) of the machine systems can be classified into typical categories. These are derived from the functional structure, and its sub-divisions as stated in TP 4 (Fig. 4), namely:

(1) *Work (transformation) organs* that exert the main working effects required to perform the technical process
(2) *Auxiliary organs* that deliver those additional effects necessary for the transformation organs (e.g. lubrication systems)
(3) *Propulsion or energy organs* that transform and deliver in the desired form the necessary energy for all other parts (prime movers, transmission linkages)
(4) *Control and automation organs* which process information and accept control commands
(5) *Connection* and *support organs* which provide internal connections of all types between the different organs. This duty includes transfer of outputs from one organ or technical sub-system to another within the machine system (e.g. energy, motion), means to provide the spatial unity of the technical system, and means to accept the supporting function, i.e. connection to the fixed system (e.g. frame or bed).

That these are *relative terms* is shown by the following example: a central lubrication system is one of the auxiliary organs of a lathe. If one regards it as a system in its own right, for delivering pressurised oil to the usage positions (e.g. bearings, gear mesh points, etc.) during the operating time, one can also recognise organs analogous to those of the complete lathe, namely work (e.g. the pump for increasing the oil pressure), propulsion (drive motor), control (valves) and connection (pipes) organs.

For the user, for whom the machine system appears as a 'black box', the most important parts of its anatomical structure are those that connect it to

TABLE 1. Categories of Machine Elements

1. General Machine Elements (ME) with mechanical modes of action:
(a) Connection elements: releasable or permanent
(b) Energy storage, mechanical: springs of all types
(c) Rotational motion elements: axles, shafts, bearings, couplings, ratchets, freewheels
(d) Transmission linkages and drive elements (mechanical):
  • wheel and roller transmission, e.g. gear wheel, worm and wheel, chain, belt, and friction drives
  • crank and slider drives
  • screw drives
  • cam drives
  Elements of transmission linkages:
  • wheels, levers, ropes, chains, belts, excentrics, cams
  • guidances, cranks, straight-line guideways
(e) brakes of various types
(f) bases, stands, frames, columns
(g) sealing elements of various types and materials
(h) lubrication elements
(i) pipe and tube elements, fittings, closure, valves
(j) containers, pressure vessels
(k) control elements

2. General ME with other functional modes:
(a) pneumatic
(b) hydraulic
(c) electric
(d) electronic
(e) biological, etc.

3. Particular ME, with designations varying according to usage area, e.g.:
(a) piston machinery elements: piston, piston ring, valve, connecting rod
(b) isolation elements of various materials and form: heat, electrical, thermal, vibration isolation
(c) lifting elements: drums, hooks, eyes, winches, rollers
(d) control and regulation elements, automation elements
(e) machine foundation elements

the environment: for the inputs, the important parts are the *receptors* as organs that transmit material, energy and information inwards to the system; for the outputs, the *effectors* with their action localities that deliver the effects outwards to the operands. It is usual for the design engineer to find these localities and their connective dimensions on the 'outline' drawing that he uses to select and apply bought-out components. An analogous 'black box' representation can be used both for the anatomical structure and for the functional structure (Fig. 4). There is no unique one-to-one correspondence (mapping) between effects (functional purposes) and a certain anatomical structure. Daily experience gives proof that a *particular effect may be attained by a number of different anatomical structures*. This statement shows that it must usually be possible to generate alternative solutions (variants) in certain steps of the design process.

3.6 ORIGINATION AND OPERATION PHASES OF TECHNICAL SYSTEMS

Origination and operation of a technical system, as a part of its life cycle, may be divided into several distinct phases and processes, which provide a clear survey of the complete *process of generating a TS*. The flow diagrams of Figs 6 & 7 serve as models for this process. If we consider the factors affecting the individual processes (operations) within this model (according to TP 6), we obtain many essential questions for setting up the design specification (list of requirements placed on the system), and for formulating individual evaluation criteria.

For instance, if one examines the operand 'human being' in the various

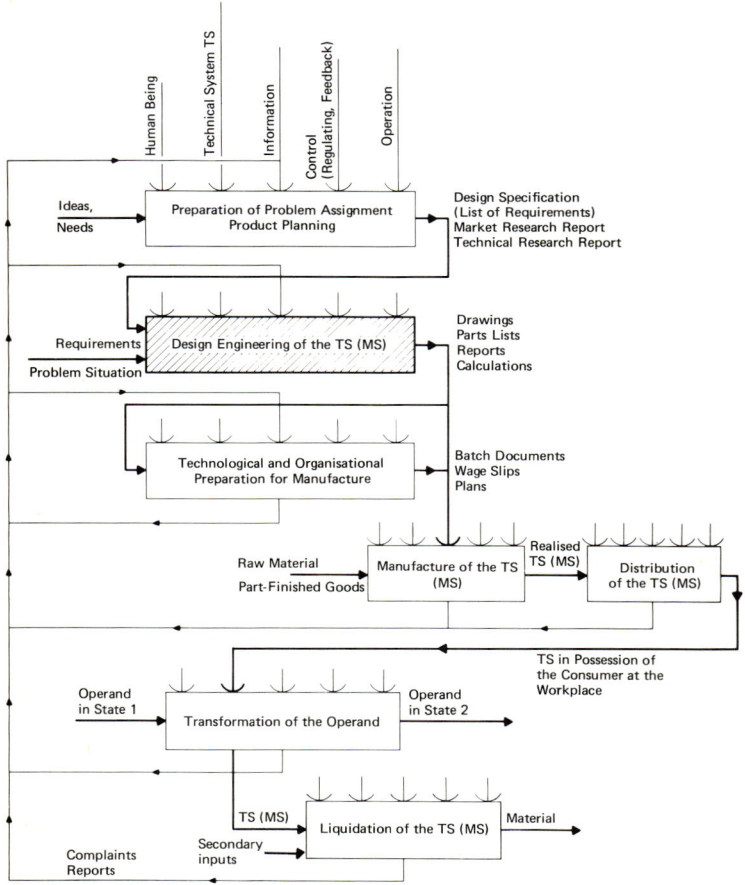

Fig. 6 Origination- and Life-Stages of Technical Systems.
   Space x, Time y

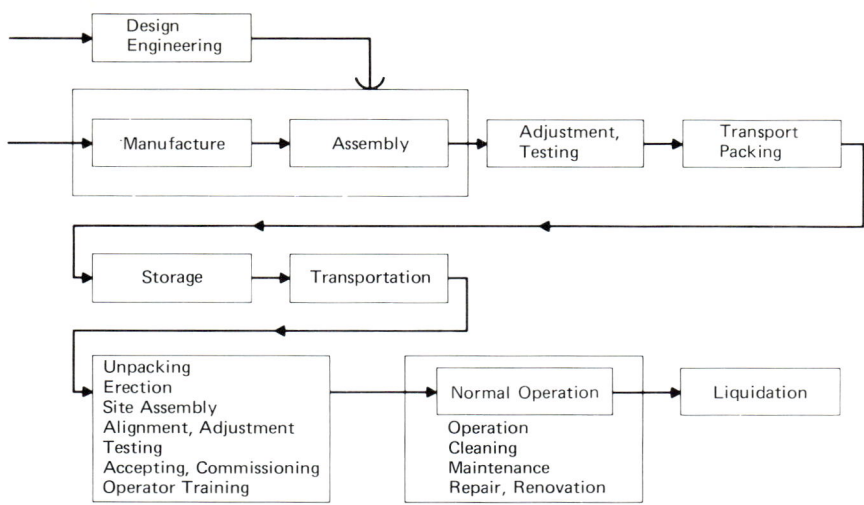

Fig. 7 Origination- and Life-Stages of Technical Systems.

processes of manufacturing and of operation of an MS (transformation process), the designer must place various requirements onto the manufacturing organisation regarding quality of workers, or on operation procedure. Consider, for example, the physical and operational differences between machine tools and agricultural machinery, which are to a large extent caused by the differing expectations of the quality of operators.

3.7 DEVELOPMENT OF TECHNICAL SYSTEMS

In the course of time, technical systems have visibly altered to a greater or lesser extent, not only in the readily apparent characteristics of form or size, but also in all other properties. For various reasons the innovation process is accelerating. If one compares a private car built in 1900 with today's model, or a calculator made in 1965 with a new one, large changes are obvious (7). The trend of keeping pace with technology always leads to gradual or intermittent improvements in the product, usually involving policy decisions at a higher management level, which are outside the scope of this book (54, 69). Some of these changes are mere *mutations* (variant designs); others can fundamentally alter the product. One then speaks of a new *generation*. The development process may also be interpreted as *changes of design properties,* of modes of action, of input values, and in the degree of mechanisation and automation.

The leading products in a particular branch of engineering represent the *state of the art,* or best available technology (in the wider sense, as referred to in Section 2.3). Misuse of this 'magic word' is possible!

3.8 ABSTRACTION AND CLASSIFICATION OF TECHNICAL SYSTEMS

The process of mental abstraction plays a large role in technology and science. The engineer usually believes that he works at a concrete level, but with many terms and expressions he moves on relatively high planes of abstraction, and these help him to create *hierarchical systems of classification* which can be of great advantage to his work. In order to observe levels of abstraction in machine systems, consider an example from the branch of machine tools (see Fig. 8). The terminology for the various levels of this systematic arrangement are taken from biology. It would have been possible to form a larger number of sub-divisions (i.e. more than six), by a finer graduation of characteristics. The arrangement shown here has among other things some advantage in clarity of representation.

By analysing the characteristics of the individual levels, and of the graphical models and representations shown in Fig. 8, one discovers *analogies* with other areas of design methodology. This comparison shows that the set of levels of abstraction is not only a characteristic for the general order exhibited by technical systems, but at the same time represents a *set of design documents* that are worked out in that sequence during methodical design work. In other words, during design one progresses through some or all of these levels from the abstract towards the concrete. A machine of the third degree of complication (a machine system, a lathe) is used as an example in Fig. 8. The same regularity exists without limitation for *all degrees of complication*. A parallel ordering of levels of abstraction of Technical Processes, as shown in Table 2, is noteworthy.

TABLE 2. Technical Processes: Classification System according to Levels of Abstraction

Level of Concretisation	Type of Process	Established Attributes	Example
0	Technical Process		
0.2	Phylum	Class of Transformation	Transformation of Material (form)
		Class of Operand	Metal
0.4	Class	Technological Principle	Dividing (reducing size, increasing number of discrete entities)
0.6	Family	Concrete Technological Principle	Metal Cutting by Chip Forming Process
0.8	Genus, Basic Process	Concrete Technological Principle	Turning
		Concrete Class of Operands	Rotational Objects
1	Operation of an MS	Totality of Parameters	Rotational Part of defined Size on Lathe SUR 5
	Behaviour of an MS		

Technical Systems

Od Operand; Wi Effect; MS Machine System; Trans Transformation of the Operand; Fu Function; M Motor

Level of Concretization	General Hierarchy			Example	
	Designation of Level	Defined by (Graphical Model)	Established Design Characteristics	Designation of Level	Graphical Representation
0	Machine System (MS)	Ms ↓ Wi (input) Od¹ → [Trans] → Od²	MS with mainly mechanical mode of action		
0.2	Phylum of MS	General Black Box Family of MS Od¹ → [Class of Trans] → Od²	− Phylum of Operands − Class of Transformations	Machine Tool	Machine Tool Rigid Body ↓ Rigid Body Form 1 [Forming Process] Form 2
0.4	Class of MS	Detailed Black Box Sketch of the Technological Principle Basic Functional Structure	− Family of Operand − Technological Principle of Transformation − Necessary Input Effects and thereby the basic Functions	Metal-cutting Lathe	Lathe Rigid Body ↓ Rigid Body Rotational Form 1 [Forming Process] Rotational Form 2 /Fu/ /Fu/ /Fu/ →
0.6	Family of MS	Detailed Functional Structure Rough Anatomical Structure Concept Sketch	− Species of Operand − Functional Structure (Functions and their Sequence) − Inputs to MS − Families of Function-carriers (Organs) − Combination and Basic Arrangement of Function-carriers	Universal Metal-cutting Lathe	Wi (output) → /Fu/ /Fu/ /Fu/ → → /Fu/ /Fu/ /Fu/ → Wi (output) (M) M
0.8	Genus of MS	Drawing of Anatomical Structure (Similarity conditions differ for different Type Series) Drawings of Common Components and Subassemblies	− Complete Parts − Arrangement − Partial Form − Some Dimensions − Types of Materials − Some Tolerances and Surface Property Definitions	Lathe type SUR	
1	Species or Serial Size	Complete Set of Workshop Documentation	Total and Definitive Specifications for Parts and Arrangement. For all Parts: Forms, Dimensions, Materials, Manufacturing Methods, Tolerances, Surface Properties	Lathe Serial SUR 5	Power x kW

Fig. 8 Technical Systems: Classification System according to Levels of Abstraction.

24   Technical Systems

The systematic arrangement developed in this section from the classifying aspect of levels of abstraction for machine systems is not the only one possible. Other aspects are total function, mode of operation, predominant manufacturing methods, complication, intricacy, etc. A comprehensive description of other classifications may be found in (8).

3.9 PARTICULAR TYPES OF MACHINE SYSTEM, BRANCHES OF MECHANICAL ENGINEERING

The levels of abstraction and their models can serve yet another purpose. Up to now we have concerned ourselves mainly with the abstraction levels of Machine Systems and Technical Systems. The statements therefore had to be formulated in very broad terms to make them generally valid. If we place ourselves in the position of a design engineer, we can relate these ideas to a particular area of expertise or branch of engineering. The designer generally works within a particular branch, and he designs a very *well defined genus* or even a particular type of machine system. In this case all statements may be re-formulated in a more concrete fashion (concretised) according to the relatively lower level of abstraction, where *many of the characteristics* (design features) *are predetermined;* for instance, instead of a general formulation of a requirement, a particular state of embodiment of a property is specified (e.g. not 'a sonic echo location system', but 'a type XYZ sub-marine sonar equipment to a requisition with a specific reference number').

Fig. 8 shows clearly the progressive concretisation of the example, where already at the first level, in the technical process, the operand and its transformation of state are established, i.e. machine tools work on pieces of material (it would even be possible to specify only metals or a very limited range of materials), and are used to change their form (especially their geometry). At the next lower (more concrete) level, only a rotational form is allowed as subject of the transformation, and this effect on the operand is based on a particular technological principle. The practical conclusion is that the *general statements can be concretised for the particular branch,* and the resulting solution possibilities can be represented in a viewable fashion on the basis of the chosen classifying aspects.

3.10 SYNOPSIS OF IMPORTANT STATEMENTS ABOUT TECHNICAL AND MACHINE SYSTEMS

MS 1: Technical Systems deliver the effects *(necessary actions)* that serve to perform the desired transformations of the operands within the technical process. Machine systems, as special cases of technical systems, use mainly mechanical modes of action to produce these work effects. In total, they increasingly tend to become hybrids (e.g. electro-mechanical systems), particularly with respect to their propulsion and control organs. The more limited term 'machine system' is therefore primarily to be regarded as a *collective term for all products of mechanical engineering.*

MS 2: The work effect is accompanied by obligatory *auxiliary, propulsion,* and *control,* as well as *connecting* and *support effects* (also termed partial functions). These constitute the classifying aspects for the functional structure of technical systems (Fig. 4).

MS 3: The *mode of action* of a technical system is formed by an active connection between a number of components that change the input quantities into output quantities (i.e. to produce the effects). Because technical systems are largely deterministic, this results in a causal chain, from causes to effects (Fig. 3).

MS 4: The effect (causal chain) is based on *natural phenomena.* Knowledge from various of the natural sciences is used, e.g. mechanics, electrics, electronics, chemistry or biology.

MS 5: The *input quantities* of technical systems are either materials, energy, or information (Fig. 3).

MS 6: Each *anatomical structure* (or technical system) is *carrier* of a well defined number of properties or property groups (Fig. 5). The individual TS differ only in the state of embodiment and magnitude (measured on an agreed scale) of the properties. In the model of the technical system (Fig. 3), all effects of the properties are shown by appropriate effect symbols (see Frontispiece).

MS 7: The *primary properties* of a machine system (those on which all others depend) are: the anatomical structure, (i.e. the machine components, with their form, size, material, manufacturing method, tolerance, and surface finish), their arrangement, and their state after assembly and adjustment. This class is termed the *design properties* (Fig. 5).

MS 8: The *quality* (total value) of technical systems may be expressed as the vector sum of individual property values. The selection of properties (criteria), their measures (scales), and the method of treatment of these values determine the validity of the evaluation statements. In the various phases of generation of the solution (Figs 6 & 7), this selection should ensure the optimum behaviour (and the optimum combination of properties) of the MS.

MS 9: The behaviour of technical systems is adversely affected by *undesired inputs* (disturbances). For this reason one should in all design phases examine the influence of potential disturbances (Figs 6 & 7).

MS 10: Technical systems cause further *secondary effects,* besides the effects resulting in the expected transformations and those due to the design properties. The danger of damaging emissions must be examined for all phases of the life cycle (Figs 6 & 7), and use of these secondary outputs (e.g. recycling) should be considered.

MS 11: Machine systems may be divided into four classes according to *degree of complication,* namely plant or equipment, machines, assemblies and parts. Each system of higher hierarchical order is composed of systems of lower order. The elements of the anatomical structure of machine systems and their hierarchy are thereby determined.

MS 12: *Machine elements* (ME) are frequently occurring structural components of machine systems. A machine element can exist as a component of the first degree of complication, (i.e. a detail part such as screw, key, split pin), or as a functionally composite group of parts (higher degree of complication, e.g. bearing, gear box, brake, coupling). In addition to the general class of ME, there exist elements that are only found in particular branches of engineering, such as a piston, valve, heat insulator, hook, see Table 1.

MS 13: The various types of effects (see MS 2) are related to the individual *organs of the machine system* which act as the realisers of the effects, i.e. work, auxiliary, propulsion, control, and connection organs that together form the complete organ structure of the MS.

MS 14: The connections between the machine systems and the environment are the *receptors and effectors,* that are also respectively the start and end of the action chains formed by the organs (Fig. 4).

MS 15: Every effect at any level of complication may be realised by a *variety of anatomical structures*. The possibilities of forming variations depend on the design characteristics—input, mode of action, and the design properties of the TS (see MS 7).

MS 16: The sequence of the phases of origination and operation of machine systems form the structure of the *complete generation* process of the MS. By analysing the factors influencing the individual phases one obtains complete information about all situations in which the MS can exist (Fig 6 & 7).

MS 17: *Development of the TS during a period of time* (innovation and further development process) leads either to small changes (variations, mutations) with basically unaltered work effects, or to larger changes when a new generation of products is created. The overall changes of properties are based on individual changes concerning:

(1) Involvement of the human being in the technical process: reduction of such involvement is accomplished by mechanisation or automation.
(2) Input quantities to the TS.
(3) Mode of action of individual action chains at various levels of complication.
(4) Individual components (structural parts), their form, dimensions, materials, manufacturing methods, surface properties or tolerances.
(5) Arrangement of structural parts.

MS 18: The *state of the art* in a branch of engineering is represented by those products whose properties have reached their highest levels of development at that time. Knowledge of this state is essential for the design engineer.

Chapter 4

Design Processes

The progressive stages of existence of technical systems are shown in a general fashion in Figs 6 & 7. These schematic diagrams also contain the design process, within which the concept of a required technical system is developed and described. One cannot assume that design has always been viewed and represented in this way, nor can one assume that it will always remain in this form, which is currently familiar to many design engineers. A group of specialist experts has formulated these models and methods of design engineering (see the bibliography), supported by fairly limited technical means.

The present state of development of design models and of their application has been reached through the centuries from a starting point of manual craft manufacture, because in earlier days all of the necessary human functions (e.g. planning, design, manufacture, distribution) could be performed by a single person, or by a small group directed by one person. From a viewpoint of the total development, today's state of affairs is an intermediate phase. Large changes are expected in this area, as is occurring in many others.

Changes in all areas of human activity are taking place under the name of 'rationalisation', with the aim of increasing economic benefit to humanity. This is also true of design engineering. We generally try to attain an optimum product in the shortest time and at a minimum cost. The latter two aims are often relegated to the background with respect to the importance of an optimal product. If important time limits are to be met, or if better conditions arise, these priorities can be reviewed, and different aims set.

4.1 BASIS OF KNOWLEDGE OF SYSTEMATIC DESIGN

The design process contains a vast range of activities, which may be justified with respect to various disciplines. We will only mention the most important sources.

Design engineering is primarily a mental activity, an activity of thinking. The *psychology of thought processes* investigates human thinking activities particularly during problem solving. A group of theories have been proposed which attempt to explain these thought processes. Without going into detail, the following are a few of the terms relevant to design method:

(a) *Association* takes place by forming connections between different concepts; the occurrence of one concept can cause another associated concept to rise into consciousness. In this way, new ideas can be stimulated by association.
(b) *Thought* can be either conscious, pre-conscious, or unconscious. The pre- and unconscious modes are sometimes regarded as irrational. One aim of methodical (systematic) design is to favour conscious thought processes.
(c) *Intuition* is understood as experiential thought in which the various stages are no longer fully conscious. In the sense of the previous paragraph, this constitutes pre-conscious thought.
(d) The causes of *thinking errors* are investigated, with particular warning against fixation (prejudice), inaccurate problem definition, and solving of problems under pressure of time.

A distinction can be made between 'rational' and 'controllable' thought. Many thought processes can be recognised as rational after the event (post hoc), although their origin was not under full conscious control. These thought modes generally can not be forced, deliberate attempts to use any thought mode or mental process do not necessarily lead to the desired results, especially if the intention is to stimulate the creative aspects. The main reason for advocating conscious thought modes is to avoid the common error of 'jumping to conclusions', without thoroughly investigating the problem. The author advises against relying on intuitive thought for present-day usage, even though this was the almost exclusive thought mode in the past.

Another general starting point for investigating thought processes is to be found in studies of *heuristics* and *creativity*. Following are the principles on which heuristic procedures (defined as 'serving to discover, proceeding by trial and error', Conc. O.D.) are based (37, 42, 49, 50, 61, 80, 91, 93):

(a) Ensure motivation,
(b) show limiting conditions (expanded, clarified problem),
(c) dissolve prejudices (no fixations),
(d) search for variants (possibilities of optimisation), and
(e) reach decisions based on evaluations of maximum objectivity (without decisions the design process is impossible).

The disciplines of *production-operations science* (including work study) have the general aim of improving the sequences and operations of human work methods. *Ten basic principles* of scientific work organisation (in a simplified form) are:

(a) Create favourable working conditions of all kinds,
(b) ensure clear formulation and understanding of every task or problem,
(c) analyse every task and its work content, and divide it into an appropriate series of partial tasks and sections; give priorities and specify time deadlines,

(d) prepare and critically assess all necessary data,
(e) choose the optimum solution for the given conditions,
(f) carefully prepare each work item in both technical and organisational aspects,
(g) prepare a plan for the execution of each larger section of work,
(h) supervise, control, and ensure proper organisation for executing the work,
(i) inspect results and compare with desired values (quality control),
(j) consider all insights and experiences, and present a written evaluation.

These principles can with considerable advantage be transferred to design engineering, regarded as a form of human work (73, 74). A good range of knowledge of design tasks is needed in order to develop concrete advice for the designer from these principles. This is particularly true with respect to item c), concerning the designer's working methods.

4.2 PREMISE FOR VALIDITY OF GENERAL DESIGN METHODOLOGY

Knowledge about design processes can be obtained by considering these questions:

(1) Are all phases of the design process *rational*, or do various irrational phases occur?
(2) Is engineering design basically a scientific or an artistic work experience?
(3) How far can one progress by considering mental *abstraction* of the objects to be designed? In other words, does a *general design procedure* (an object-independent method) exist, or are there only particular design methods (e.g. for the design of cranes)?

These questions are formulated in a rather 'black or white' fashion, but they can really only be answered in 'shades of grey' between the extremes. Any design engineer will acquire generalised procedures, usually by learning from simple problems, before tackling more complicated ones. An additional question (particularly relevant to engineering education) is therefore:

(4) Can such design procedures be acquired only by (unconscious) experience, or can a generalised procedure for design be (consciously) learned as a visible and applicable process?

Again, the indications are that the answer lies somewhere between these extremes (52). Nevertheless, recent investigations have shown that much more can be *learned* than was previously thought possible.

A positive answer for the rational viewpoint, that recognises a general design method independent of particular objects to be designed, has provided the starting point for the research into design methodology that has taken place in the past twenty years. Intuition in design work has also received its appropriate position (12).

4.3 TASKS AND METHODS OF DESIGN RESEARCH

Design research, particularly in the field of working methods, raises the questions:

(1) Into which *components* (design operations) can an object-independent design process be sub-divided or decomposed (question aimed at determining structural elements in the design process)?
(2) Which *sequence* of design steps permits attainment of an optimal technical system, reliably (with a minimum of failures), and by the shortest path from the assigned problem statement (question aimed at determining a general procedural model for the design process)?
(3) Which *modifications* of the general procedural model arise from changes in factors affecting the design process?
(4) Which *tactics* should be selected for individual design operations to obtain optimal results?

In addition, design research must also attempt to answer questions about *interdependence* and contingencies within the *system* (the design process) that may be important for planning, control and other considerations. The knowledge that has been collected about working methods can play important roles in evaluating the proposed design methods, and comes from a variety of sources:

(a) Good designers are *observed* and the resulting observations generalised (e.g. 72). Influences of the type of designer, the engineering branch, and the form of problem statements on such observations must be recognised.
(b) *Activity analogies* in other areas of knowledge are sought, and positive results transferred. The analogy can be broad or narrow, e.g. there are some analogies between problem solving methods in mathematics and design (84, 96).
(c) *Theoretical research* can thoroughly investigate the areas of design, can find the influencing factors and their interrelations, and can propose and test hypotheses. Results from other areas of knowledge (sciences, arts and humanities) are brought to bear on this endeavour.
(d) Knowledge can also be transferred *directly from other areas of knowledge* into design methods, e.g. from psychology (about thought, behaviour, problem solving), and heuristics.

This is obviously not a one-way process, knowledge of 'what works' in performing a design process has stimulated a number of investigations in psychology and in other areas of knowledge.

Research into methods of design engineering cannot build a science in isolation. Design work is critically dependent on many other areas of knowledge, because it is largely an information generating and processing activity. The design engineer must therefore be concerned with all the 'hard' engineering research, development and experience, and with other sciences and their developments, because he has a major duty of avoiding catastrophic failures. Nevertheless, it would be very wrong to assume that it is unnecessary for the engineer to study design.

This area of engineering education and practice is slowly receiving the attention it deserves, as a unifying influence for all engineering investigations and activities. As our knowledge of the physical world increases, so a better knowledge of the human mental world becomes more important, especially for those creative activities that exert such a wide influence on human life as design engineering.

4.4 GENERAL MODEL OF THE DESIGN PROCESS

Any brief representation showing the problem area covered by the term 'design processes' should appear in the form of a system, and should show relationships within the design process, by means of a generalised model. The model in Fig. 9, permits these interpretations:

(1) it defines engineering design as a process of *transforming information* from a customer's statement of requirements to a full description of the proposed technical system
(2) it shows a basic *structure* for the process
(3) the effects (actions) that are exerted onto the design process (viewed as a technical process analogous to Fig. 1) are brought to bear by the *designer and by his working means* (as influencing factors affecting the process)
(4) apart from the designer and his working means, *various other factors* affect the transformations and their economics, namely,
   (a) Available technical information, including specialist design information
   (b) working methods used,

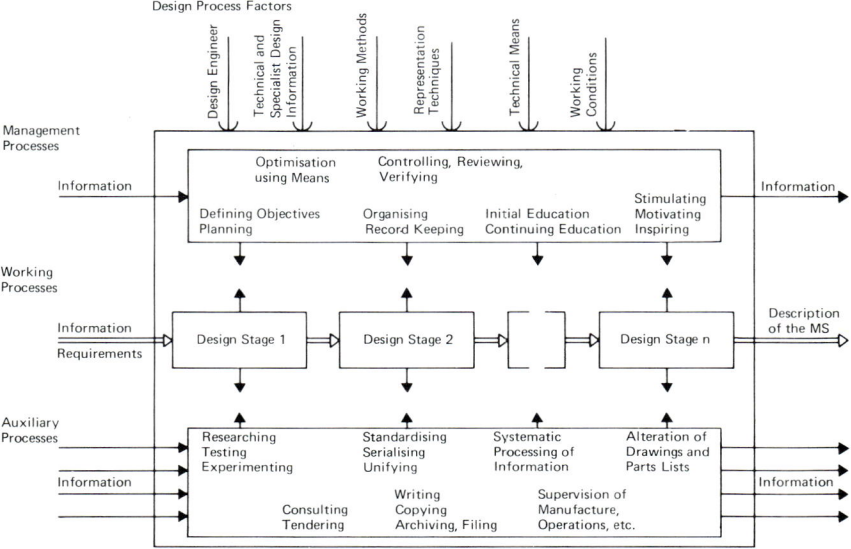

Fig. 9 General Model of the Design Process.

32 Design Processes

(c) techniques of representation employed,
(d) quality of management, both of the design process and of the design personnel,
(e) working conditions and environment.

Large differences in the duties assigned to professional engineers exist between various branches of engineering. In some companies, the engineer is involved in the early process of defining the initial problem statements (assigned requirements, involving negotiations with the customer), and therefore design research in some countries has also concerned itself with that step. In other companies, the design engineer does not need to be so concerned with the phases of problem definition at such an abstract level, especially if more persons with professional engineering qualifications and experience are employed in upper management levels, including customer relations, sales, and at directorial or board levels. Other differences can be seen in the way that a problem is assigned. On many occasions the assigned problem is set by the engineering designer for himself, but the better documented cases are those in which the problem is assigned to the designer by an agent external to the design process as such.

4.5 STRUCTURE OF THE DESIGN PROCESS

Regarding the structural elements of the design process, an appropriate investigation can provide important pointers to enable us to construct a procedural model, which uses parallels to the knowledge of TS and MS developed in the previous two chapters. Ordering and surveying the complicated observations and terminology of design activities can yield two useful results:

(1) A hierarchy of *activities* in the design process, ordered according to their complexity (complication and intricacy),
(2) *Activity blocks* that represent recurring logical connections of some of the activities to certain strategic aims.

A hierarchical arrangement of most of the design activities is shown in Fig. 10. The activities shown at each of the levels in this scheme are intended to extend the concepts relating to activities within the design process, and are at the same time components of each of the activities in the higher levels. Consequently all higher activities are composed of a mixture of all the lower activities ordered in their respective classes.

Considering the third level of Fig. 10—the basic operations—its position in the hierarchical order may be seen, but this level also shows a cyclic *block of operations* that constantly recurs during problem solving. The block scheme shown here is identical to the procedural model of problem solving developed in the next chapter. It may be used as the strategic concept of the whole design process, or of its parts, e.g. (1.1) conceptualising, (2.6) form design, (2.9) dimensioning, etc. The basic operations shown in this block of operations are among the most frequently used general activities in the

Design Processes

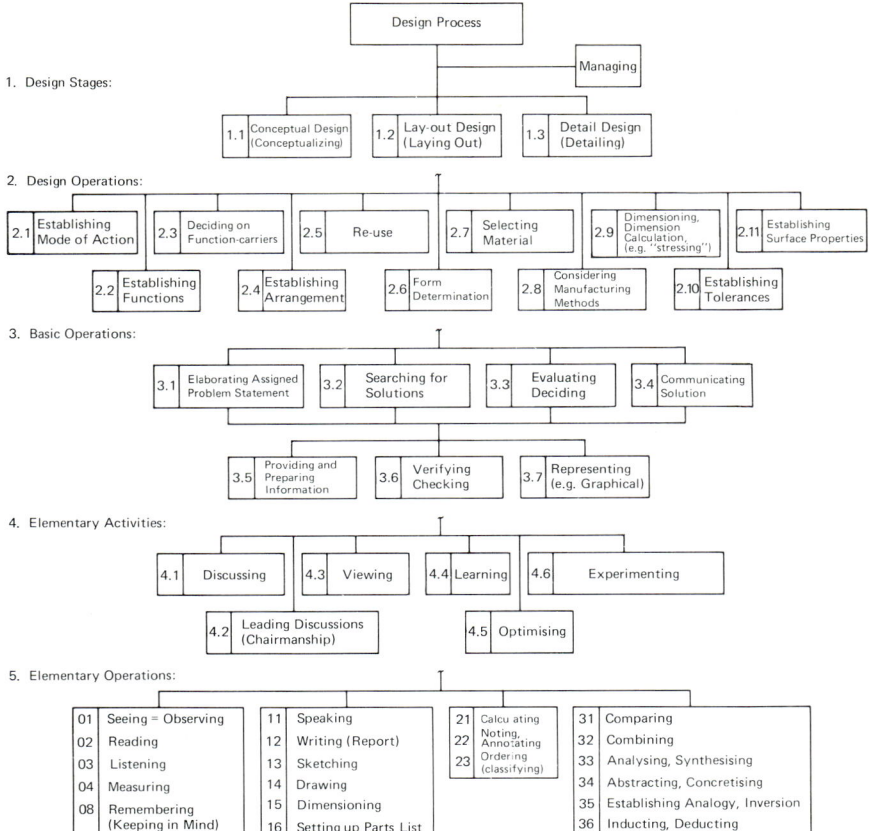

Fig. 10 Structural Parts of the Design Process.
(Hierarchy of externally observable human operations, with directly verifiable output in the form of recorded material)
NOTE: 'thinking' as interpreted in this figure is a non-observable human activity, but is essential to all design tasks. In order to make this diagram more complete, a box labelled '4.7 Creative Thinking' could be included at Level 4.

design process (compare 94), and should be mastered by all design engineers:

(3.1) Elaborating the Assigned Specification into a full Design Specification (Problem Statement), including development of evaluation criteria for the future solution
(3.2) Searching for Solutions
(3.3) Evaluating and Deciding
(3.4) Communicating
(3.5) Providing and Preparing Information
(3.6) Verifying and Checking
(3.7) Representing

Among the important *strategic manoeuvres* in the design process are:

(1) *Iteration*: which is used where a direct solution is not possible. In a procedure similar to that used in mathematics for the approximate solution of a system of equations, one makes certain assumptions in order to proceed towards (e.g. calculate) a solution. In the next stage, these results when used as improved assumptions help to determine a more accurate solution (values), and if convergence is sufficiently rapid, a few cycles (iterations) deliver the appropriate solution.

(2) *Abstraction* (to progress towards completeness): in this process, one initially ignores unimportant aspects, and concentrates on the important and decisive ones. This method permits easier entry into a design problem, by making the designer concentrate on realising the main working effects.

(3) *Concretisation*: leads from rough preliminary solutions, towards definitive, more finely tuned ones. In practice, this strategy merges with others.

(4) *Improvement*: starting from an existing solution that is judged to be capable of improvement, a satisfactory solution can be achieved by criticism and modification.

(5) *Strategy of the Problem Axis*: a forward progression from the problem (or its symptoms) towards physical means may reach a stage that allows one to achieve a solution that is merely palliative, or that paralyse further progress towards a viable solution. Reversal of the search direction towards the causes of the problem can also be attempted ('go back one step'). In some cases, this procedure can permit the designer to 'get rid of the problem' by avoiding it from a more abstract starting level.

4.6 DESIGN STRATEGY, PROCEDURAL MODEL

The predominant centre of interest in recent design research has been the generation of general procedural models. This may be seen from the many models that have been published. It is important to note that these were produced in many countries, in various branches of industry, at various places of work, with different working conditions, and are based on varying objectives and assumptions, in various forms and with their own terminological apparatus. We do not wish to discuss details of the individual 'schools of thought', a different publication (9) had that aim. The general model introduced in the next section is based on the current state of knowledge, and attempts to integrate the positive experiences of some of the existing models.

In the spirit of the explanations of technical systems (statement MS 7) we are searching during the design process for an *anatomical structure* that will be the *carrier of the desired properties*. This structure should then be accurately described by the appropriate elementary design properties. The relationship between the given requirements (as input to the design process) and the design properties to be found (as output from design) is complicated, firstly because the number of these relationships is usually

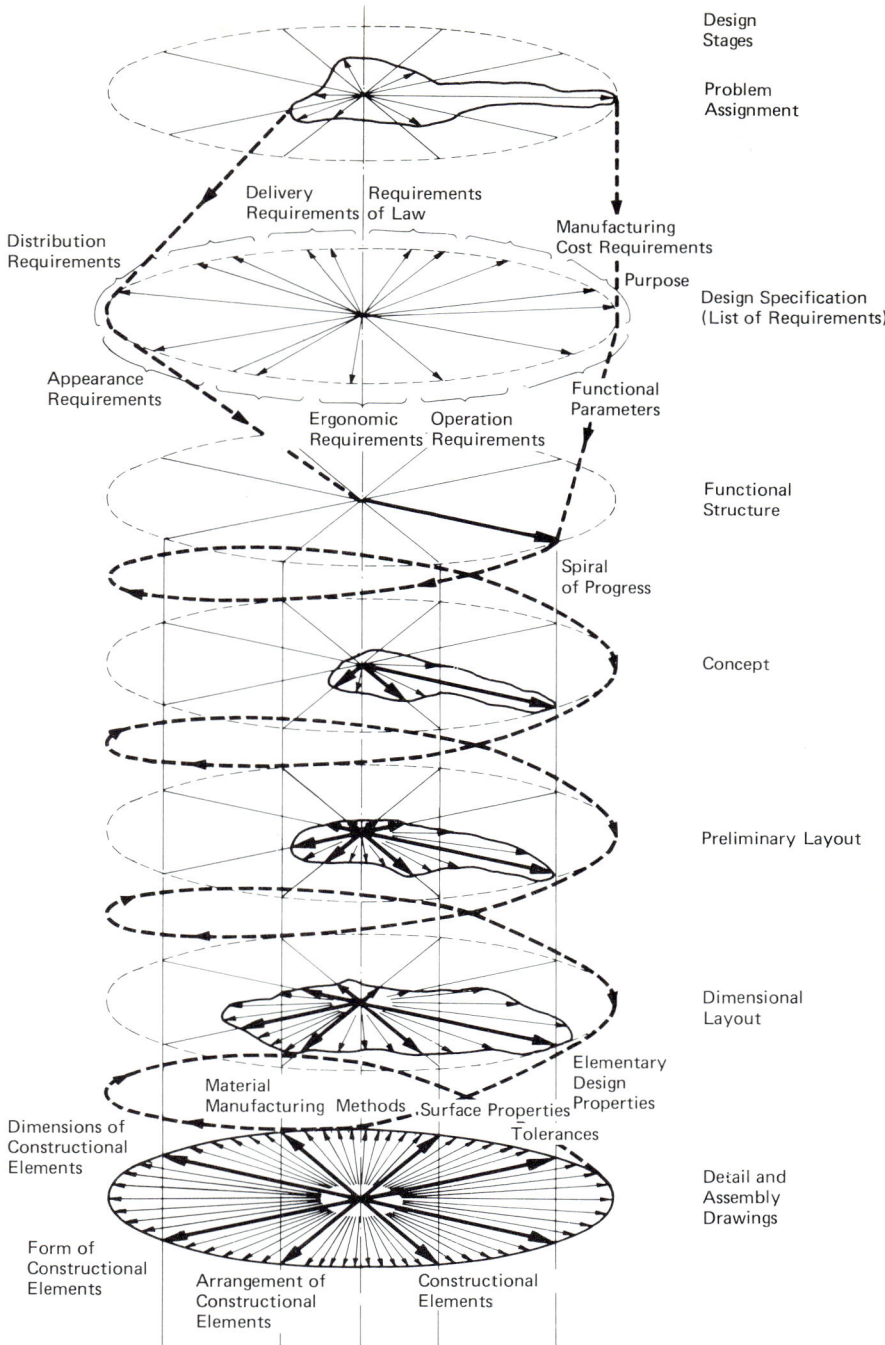

Fig. 11 Application of the Degrees of Completeness and of Finality of the TS-Properties during Design Progress.
NOTE: The notional circle drawn at each level represents 100% fulfilment of the required properties at that level, i.e. completeness of information on requirements at level 2, finality of TS description at level 7 (see also (87)).

large (compare Fig. 5), secondly because quantitative knowledge about some of the relationships is lacking, and thirdly because all phases of the design process are mutually interrelated.

A direct transition from the list of requirements to a hypothetical concrete anatomical structure is generally inconceivable. One must initially attempt to realise the most important properties in an *iterative procedure*, and then investigate the resulting anatomical structure to determine the adequacy of other properties. For such a procedure to result economically in an optimal solution, it is necessary to form and refine the anatomical structure in a number of steps, for example:

> 1st step: abstract anatomical structures, incomplete and preliminary physical concept, organ structure well defined from a previously established functional structure
> 2nd step: partially concretised anatomical structure, complete and definitive preliminary layout
> 3rd step: fairly concrete anatomical structure, fairly complete dimensional layout
> 4th step: concrete anatomical structure, complete and definitive dimensional layout.

The concreteness (completeness, fidelity) achieved in the resulting anatomical structure depends on the design engineer establishing the design characteristics (features), including the elementary design properties. The larger the number of established design properties, and the more completely they are determined, the more concrete is the resulting anatomical structure. Fig. 11 shows this relationship in graphical form (compare 87).

Evaluation (including that of form) is not without its problems. The more abstract the anatomical structure, the more difficult it is to judge and verify conformity to the requirements. Evaluations and decisions based on experience and apparently reasonable assumptions can therefore sometimes be wrong, although this should be rather the exception, because in most cases the total quality of a potential solution can be estimated with sufficient accuracy from a few of the available criteria. If the uncertainty is too large, it is recommended that a definite decision should not be made at this more abstract level, but that a further concretisation is performed before coming to a decision. Further theoretical statements and discussions may be found in the literature (2, 5, 6, 9, 17, 22, 25).

4.7 FITTING THE GENERAL MODEL TO PARTICULAR CONDITIONS

The *ideal procedural model* is always developed on the basis of certain assumptions. A plan for solving a concrete problem under concrete boundary conditions must take on a particular form. The factors causing such mutations are:

(a) Regarding the *technical system*: its degree of complication, degree of originality (possible variants), number and degree of difficulty of requirements.

(b) Regarding the *design process*: state of the influencing factors—quality of designers, state and availability of technical information, working means available to the design team (see Section 4.10), management of the design process, working conditions.
(c) Regarding *production*: number of parts (per machine system, and for each detail part *vs.* type of production method), time deadlines, experimental and manufacturing facilities, traditions, organisation.
(d) From *society at large*: standards, regulations, environmental protection and other restrictions.

In view of these factors, which clearly influence the procedural plan, one must distinguish between the following concepts:

(1) *The procedural model*, which gives information about the ideal flow of work (sequence of important phases) which should be followed during the design process, either in verbal form, or in graphical form as a flow diagram. The model usually views only the technical and logical relationships, assuming an *ideal* definition for the state of the various factors, such as complexity of product, quality of designers, technical information, deadlines, management, working conditions, etc. The validity (applicability) of the model is otherwise usually exceptionally large.
(2) *The procedural plan*, which is worked out for a particular problem, in contrast to the generally valid procedural model. Consequently all given factors may be drawn into consideration. It is derived from the general, theoretically based procedural model, and is fitted to the given circumstances, especially with respect to time deadlines. It is possible and in many ways better to represent this plan in the form of a critical path network.
(3) *The procedural manner*, which constitutes the ways and means of working of an individual design engineer. It is valid only for the individual designer and is strongly influenced by his personal qualities and practical experience. It reacts clearly on the form of the procedural plan.

4.8 DESIGN TACTICS, METHODS AND WORKING PRINCIPLES

The design operations, as elements of the design procedure, are often relatively complicated. As with the whole process, the individual operations demand a *planned procedure* to attain a defined objective under the given conditions. We are concerned with rules for behaviour in carrying out usually small tasks with relatively concrete problem formulation, that (by analogy) we term design tactics. The tactical instructions are found in the methods and working principles developed for design procedures.

Methods

Delimiting the term '*method*' is not easy. On the one hand the word is used to describe large activity complexes, e.g. modelling techniques, market or value analysis (and these also contain their own detail methods); on the other hand 'method' is used for more limited procedural instructions that can also be regarded as working principles within Section B of Table 4, e.g. the method of stepping backwards (see Table 3), the Descartes method, or aggregation. Similar to the above way of fitting procedures to particular circumstances, we will distinguish:

(1) A *method*, which is a system of methodical rules that define classes of possible procedures that should lead from a given starting position to a desired final situation.
(2) The existence of a method permits the creation of a *plan* to define the sequence of activities for a particular case. Each method allows creation of a large number of such plans.

TABLE 3. Design Tactics: Survey of Methods
Comments: some methods are listed as basic principles, because the boundary between these two categories is indistinct. The methods are named by reference to commonly used terms (see also (65)).

Name of Method	Characteristics	Objectives
Aggregation	Combination of machine sub-systems into a single system, or of functions of a number of organs into one organ	New properties Simplified structure
Adaptation	Modifying or partial transformation of an existing MS for different conditions	Reliable solution for new conditions
Application	Application of an existing MS for new functions	Application of proven MS to new areas of use
Dimensional Investigation	Technical and economic properties of the MS are brought into a mathematical relationship and extreme values found	Find optimal solution
Evaluation	Find technical and economic valuation by point-counting	Find best variant among a few
Brainstorming	Collect ideas in free discussion without criticism	Find many solutions to a problem
Descartes	Four Principles: criticism, division, ordering, create overview	Correctness and effectiveness of the thought process
Analysis of Properties (Attribute Listing)	Thorough analysis of every property of the MS	Improvement of the existing MS
Method of Invention	Procedure of invention, applied to design work	Find new solutions
Systematic Search of Field	Research all directions starting from fixed points of the region	Obtain completest possible information
Questioning	By applying a system of questions, find gapless information or produce mental stimulation (e.g. six work-study questions)	Obtain completest possible information
Mental Experiment	Observe an idealised mental model at work	Testing of an idea, determination of behaviour
Iteration	Starting from assumed values, obtain progressively closer approximation of all values	Solution of a system with complicated interactions
Incubation	After thorough preparation of the problem, take a break	Find solutions by intuition

(3) A *personal working mode* is based on the applied methods and is suitably adapted for the individual designer (usually for himself).

According to the objective to be achieved in that phase of work, methods may be grouped into those for finding solutions, for evaluating, for determining properties, for eliminating thought errors, and others. An important criterion for distinguishing between methods is the ability or property of the human being that the method addresses or stimulates. For instance, brainstorming and synectics use the phenomenon of association; systems thinking provides a guide towards planned procedures and order; and questioning brings about a consciously discursive thought mode. A few important methods are listed in Table 3 (49, 59, 65).

Table 3 (*continued*)

Combinations with Interactions	Combining of MS or of properties to obtain new and higher (more complicated) effects	Derive new solutions from existing MS
Systems Approach	Systematic working in every situation requiring a solution or decision	As far as possible complete investigation of an area
Market Research	Systematic collection and classification of market information	Establishing marketing conditions
Modelling Technique	Representation of the TS for various purposes	Determination of behaviour and other properties of the TS
Morphological Matrix	Enumeration of function-carriers to solve partial functions, in matrix form	New solutions by combinations of function-carriers
Critical Path Network	Graphic representation of activities and their duration	Create an overview of sequence and timing and find the critical path
Experimentation	By measuring and testing, obtain desired values	Determination of the properties of the MS
Synectics	Team analyses problem and searches for new solutions through analogies	Discover new solutions
Technico-Economic Design	By technical and economic evaluation, find and improve the strong features of the design	Determine one best solution from among a number
Step Forwards/Backwards	Attempt both solution directions, from 'is' to 'should be' and reverse	Find the most favourable path to a solution
Value Analysis or Engineering	Analysis and criticism of the existing solution from the viewpoint of economics	Improvement of the economic properties of the MS
Division of Totality	Tactical procedure based on division of a whole concept or problem into component parts	Create overview, generate partial solutions
Methodical Doubt (Scientific Scepticism)	By systematic negation of existing solutions, search for new solution paths	Find new solutions
Method '6-3-5'	6 participants, each write down 3 ideas within 5 minutes, then pass on to next person for similar 3 ideas	Find many solutions

TABLE 4. Design Tactics: Survey of Working Principles for Design Engineers

A. General Principles
1. Critical acceptance of all given information: do not accept any information without examination and verification
2. Control principle: every result must be examined. Use an advantageous control strategy and tactics, and examine important requirements, e.g. function, realisability, economics
3. Principle of effectiveness: in every process one should strive for maximum effectiveness
4. Principle of economy: economy is the first requirement for design. The function of the MS should be attained in the cheapest fashion. Depending on company policy, and on ethical considerations for the design engineer, this 'cheapest fashion' could be cheapest first cost, or cheapest running cost, or preferably cheapest whole-life cost
5. Optimisation principle: aim for optimum solution or best compromise solution for the given conditions, in optimum design time, with optimum care and accuracy
6. System and totality principle: every object and every process is simultaneously a system and a system element in a larger system. All connection, interactions and relationships should be considered
7. Principle of recording information: human memory is unreliable. Every important item of information should be recorded (e.g. written down) and classified (catalogued) in an economic fashion
8. Ordering principle: every area of knowledge should be classified (e.g. Dewey decimal classification system)
9. Overview principle: create a usable and comprehensive survey
10. Principle of methodical and planned procedure: guide the progress of activities in a methodical and planned way. Use at least one solution path, and a different control or checking path

B. General Principles During Search for Solutions
1. Orientation principles: carefully determine the state of the art in the relevant area. It may be useful to attempt to think of possible solutions to the problem, using intuition and imagination, before applying this orientation principle. The search for the state of the art can then proceed with better knowledge of the questions relevant to the specific problem, and with less danger of mental set, or fixation; but consider also the opposing danger of 'jumping to a conclusion', and defaulting on this orientation principle
2. Accept good existing solutions, but only after critical examination
3. Principle of accurate problem formulation: for each sub-division of the problem, formulate each step of the problem assignment
4. Principle of abstraction: formulate abstractions from the existing concrete circumstances to find new paths

Working principles

Besides the methods, the designer has at his disposal another basic tactical tool: working principles. They give, in a short form, important general instructions for appropriate *behaviour* for the designer to use in the particular working situations. Various selected basic principles are listed in Table 4. Some of these may be regarded as requirements for the designer, others present relevant items of technical information, or guidance for management of the design engineer's work.

It is recommended that each design engineer should produce such a list from his own experience, and thereby to progressively create for himself an *important personal working aid*, a procedural check-list.

4.9 REPRESENTATION DURING DESIGN

The ways and means of representing more or less concrete artefacts for the purpose of communication (with oneself and others) plays such an

Table 4 (*continued*)

5. Division principle: divide every complex problem into sensible sub-divisions
6. Variations-combinations principle: combine suitable elements into a totality, vary the elements that can perform equivalent functions
7. Incubation principle: permit the sub-conscious to work. Alternate productive phases with rest periods

C. Principles Concerning Quality of the TS
1. Design with respect for function: then 'design follows function' (Walter Gropius), i.e. if it *is* right, it also *looks* right, but not necessarily the converse
2. Design for the market: fulfil all customer requirements
3. Design for operation: consider operational safety and human operator conditions, minimum space consumption, minimum dimensions
4. Design for the human being: offer maximum protection for the human, avoidance of difficult or monotonous human work, aim for minimal human fatigue
5. Design for appearance: keep in mind the aesthetic effect of the product
6. Design for packaging, storage and transport: create favourable conditions
7. Design in accordance with regulations: take into account all standards, codes, laws. Do not copy products protected by patents, trade marks, design registration, etc.
8. Design for manufacturability: achieve the most economic realisability of the product with available manufacturing systems, aim for optimal manufacturing methods
9. Design for ease of assembly and maintenance: examine assembly methods for the TS, avoid special tooling
10. Aim for minimum manufacturing and running costs
11. Design for adequate strength: ensure appropriate strength and stiffness of the TS, considering also fatigue, creep, fracture mechanics, resonances, wear, and other modes of failure or deterioration
12. Design against corrosion: ensure appropriate resistance to corrosion for the TS
13. Consider thermal expansion of the system and its elements (steady-state and transient)
14. Provide adequate lubrication
15. Professionalism in design: simple structure, simple form, optimal dimensions, suitable materials, surface structure as coarse as permissible, largest possible tolerances

D. Principles Concerning Representation Techniques
1. Aim for clear, complete, unique and unambiguous representation of the TS
2. Economic, optimal representation
3. Purposeful representation (considering the receiver of the communication)
4. Consider manipulation, handling and archiving (filing) of documentation

---

important role in the design process that it features as one of the influencing factors in design (see Fig. 9). It is by no means sufficient to think only of technical drawing, which is a two-dimensional graphic abstraction of a real object; other abstractive representations (coding or mapping of a real artefact into verbal, graphical, mathematical, or computer analogues) place continually increasing demands on the designer, particularly with respect to the various functions of representation: describing, fixing, aiding visualisation and communication. When the economics of design work are considered in this context, further demands arise.

A survey of the items of *basic design documentation* and their *information content* is shown in Fig. 12. The representations used within this book show some of the application possibilities (particularly in Chapter 6). For further details see the literature (1, 13, 28, 85, 90).

4.10 WORKING MEANS IN THE DESIGN PROCESS

Every object, apparatus or system that aids the designer during his work is termed a *working means*. This area of working means is very wide—ranging

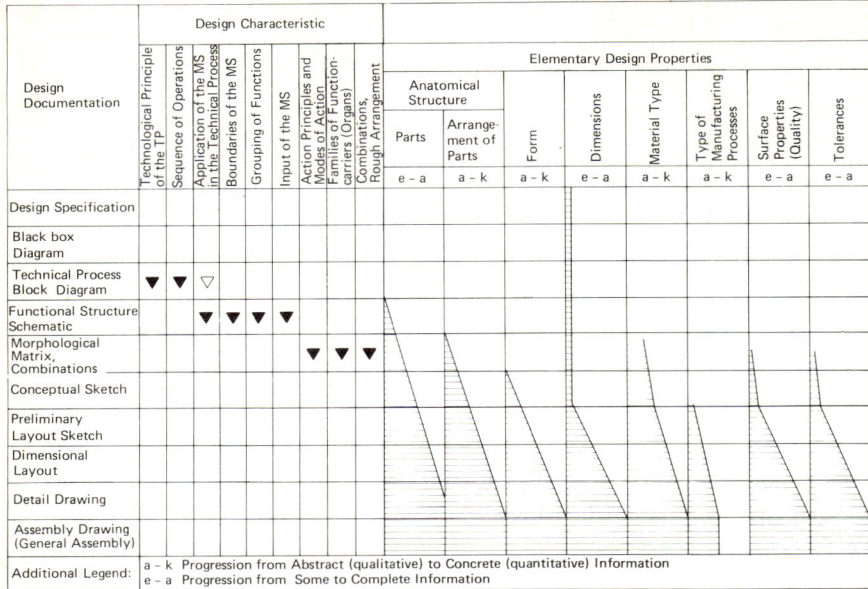

Fig. 12 Survey and Contents of Design Documentation.

from pencil and paper to sophisticated computers. The most advanced stage using electronic data processing (usually known as CAD, computer aided design), shows great promise, especially in the recent developments which bridge into computer-aided manufacture (as CAD/CAM) and numerical control of manufacturing equipment. Details may be found in the relevant literature (31, 38, 43).

4.11 SYNOPSIS OF DESIGN PROCESS STATEMENTS

(The German original uses the abbreviation KoP—*Ko*nstruktions *P*rozess—for referring to these Design Process (DesP) statements.)

DesP 1: The design process is an element in the process of generating technical systems. In the design process, the starting information (in the form of requirements or descriptions of the assigned problem situation) are transformed into the *visualisations and descriptions of the desired TS*. Description of the technical system is performed by means of elementary design properties (MS 7). Consequently the search for suitable design properties constitutes the main activity of design work.

DesP 2: As a technical process, the design process (DesP) may be divided (according to TP 4) into a *finite number of partial processes* (operations) (Fig. 10). The majority of these belong to a few recurring classes of such operations, regardless of the type of design process in question.

DesP 3: *None* of the operational phases of the design process is regarded as *irrational*. The design process is composed of rational operations.

DesP 4: The design process can be *neutral with respect to objects* to be designed. A model design process exists that may be applied to the design of *all kinds of MS* (and of TS).

DesP 5: As for other types of process (TP 1 and 6), the quality of results, the duration of the process, and economics, are influenced by a *system of design factors*. These include not only the assigned problem statement (as input to the design process), but also:

(a) The design engineer (his professional profile and his personal characteristics),
(b) technical means (working means and their use),
(c) state of knowledge (available information), particularly technical information, knowledge of working methods, and methods and techniques of representation,
(d) control of the process (management), and
(e) environmental conditions (time and space in which the occurrences take place).

DesP 6: The methodical form of design work is founded on relevant *insights from psychology, heuristics, and other areas of knowledge*. The relevant principles of these sciences may usefully be employed in individual cases.

DesP 7: The complicated relationships between the desired properties of the technical system (the requirements) and the desired elementary design properties forces the design engineer to use a repeated *iteration procedure* based on the *progressive concretisation* of the design work, i.e. a movement from incomplete to complete information, and from approximate to definitive values.

DesP 8: Design work demands various *directions of progress* to establish the states of the technical system. The basic movement in this activity is towards the search for an anatomical structure to realise the required output effects (purpose function). This search progresses through a number of iterations on each of a number of abstraction levels. The general model in this book considers that an optimal procedure consists of four such iterative cycles (Fig. 11).

DesP 9: The strategy of methodical, systematic and planned procedure in design is generally defined in a *procedural model* that has been set up for an assumed state of the design factors (DesP 5) consisting usually of very generalised conditions. The concrete conditions which arise from a specific assigned problem and for a particular design engineer lead to mutations of the model, to yield *procedural plans*, or personal working modes, etc. The plan for the design process should be *as highly structured as possible*, the details reaching into 'design operations' or 'steps' where applicable.

DesP 10: The designer's behaviour in the design operations can be controlled by means of suitable *methods* or *working principles*. The relevant area of knowledge is that of design tactics.

DesP 11: *Representation* in design work has high priority if the solution process is to be successful.

Chapter 5

General Model of Methodical Procedure During Design— The General Procedural Model

The *starting situation* for formulating a general procedural model has already been described in the previous chapter:

(1) Many procedural models exist, but in spite of certain similarities, they contain very variable information that provides few possibilities for the practitioner to make useful comparisons.
(2) In practice, and in the majority of cases, work still proceeds along traditional paths, i.e. using intuitive steps.

5.1 REQUIREMENTS OF THE PROCEDURAL MODEL

The requirements for the model may be summarised as follows:

(1) The design process should be *structured* into *easily recognisable, conscious solving steps* (to ensure transparency, and use of discursive procedure) in order to reach a near optimal solution in the most economic way (according to the principle of efficiency, Table 4).
(2) The model should as far as possible be *object neutral* (independent of the object to be designed), and be capable of concretisation for particular product families.
(3) The model should be formulated for *clearly defined factors* (assumptions). The influence of these factors leads to mutations of the general procedure.
(4) Clear *relationships* to other areas of design science should exist.
(5) *Substantiation* of each step should be possible by reference to insights from other areas of knowledge.
(6) The *results* of other models must be respected.
(7) An understandable *form* for a practicing design engineer should be achieved.
(8) *Application areas* for the model should include: practicing engineering

designers, team work, management, planning, organisation, basis for CAD, teaching, introduction of sensible formalisms that should serve to rationalise design work.

5.2 CONCEPTS OF THE MODEL

The concepts on which the procedural model (Fig. 13) is primarily based are:

(a) *Patterns* recognised in the areas of *technical processes* and *systems* (Statements TP and MS, Sections 2.10 and 3.10). The basic assumptions are:
- The TS is regarded as the highest abstraction of the object (operand abstraction).
- Design is the search for a suitable anatomical structure, which is described by elementary design properties.
- The intermediate states of development of the TS are the important points in the procedural model (Fig. 14).
- Solving for the total function or for the effects proceeds stepwise through various levels of complexity (i.e. not by direct decomposition into elementary basic functions).

(b) *Insights from psychology, heuristics, work study sciences* that are adapted for design work (see statement DesP 6). The basic assumptions are:
- Concretising of the anatomical structure occurs in *four cycles* (Fig. 11).
- The system of *verification* based on evaluation is composed of three basic steps: review (control, and checking) of work performed in each design step and after each activity (not shown in the diagram of the model), review after each major design phase (5 verifications), and review before release. (This final review is shown in the model after completion of the layout. A similar review can be performed after conceptual design, and another must be performed before manufacturing planning, i.e. at the end of the illustrated procedural model.)

The general procedural model shown in Fig. 13 is set up assuming these *conditions*:

(1) A *new development* with a technical system of the degree of complication 'machine', 'apparatus', 'equipment', or 'plant'. For detailing, design phase II is omitted. For adaptive design or further development of existing TS other phases or steps become redundant.
(2) *Sufficient design time* is available (no pressure of deadlines),
(3) *One-off or small-batch production*. For large batch and mass manufacture, phase IV changes to 'elaboration of complicated systems', and is augmented by steps of prototype manufacture and testing, and where needed design of mass-manufacture methods, machinery, jigs and fixtures, tooling, etc.
(4) An *average design engineer*, and average *state of information, working means, general conditions* and *management*.

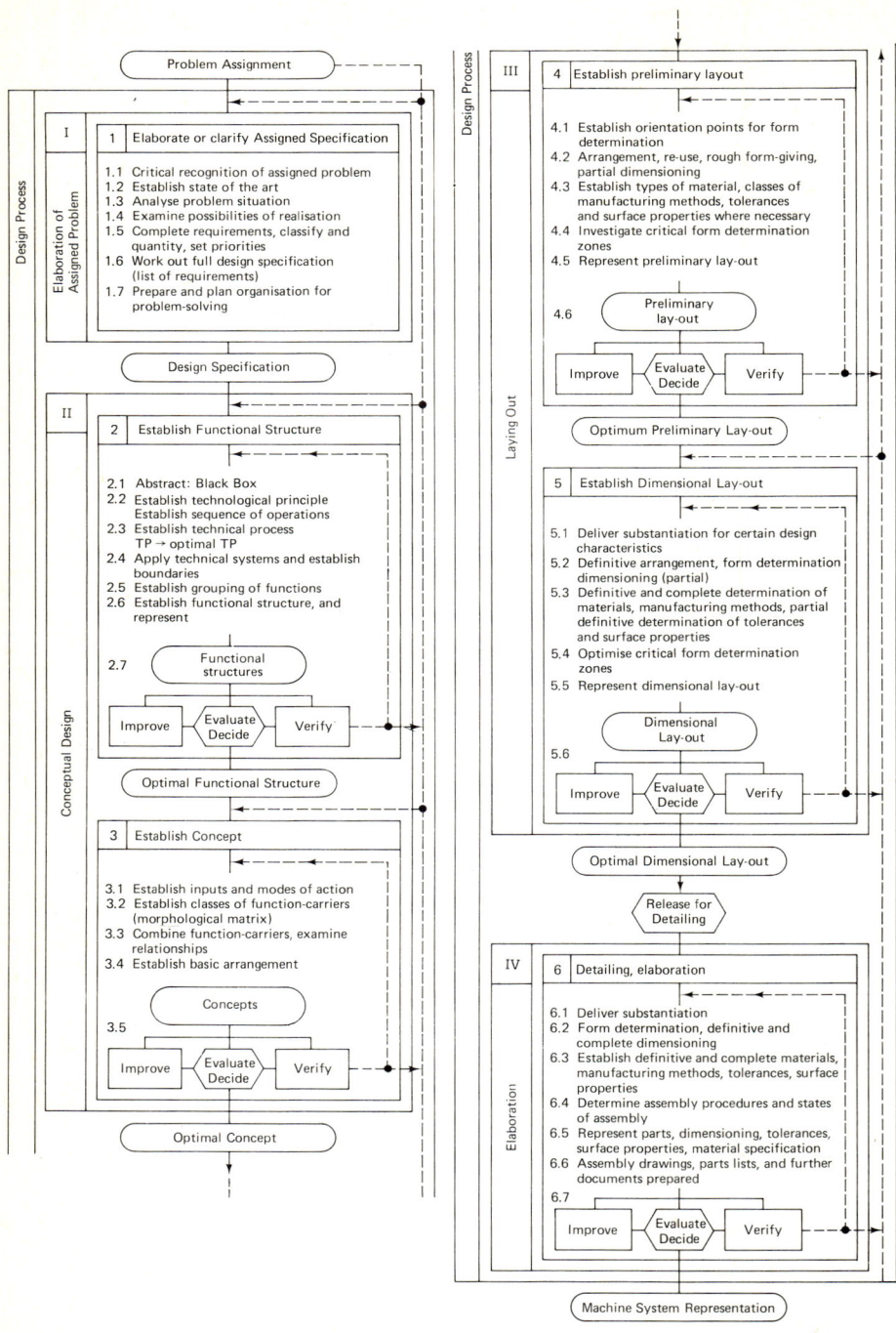

Fig. 13 General Procedural Model of the Design Process.
NOTE: regarding items 1.2, 1.3, and 1.4 see also note in Table 4, section B, item 1.

General Model of Methodical Procedure During Design   47

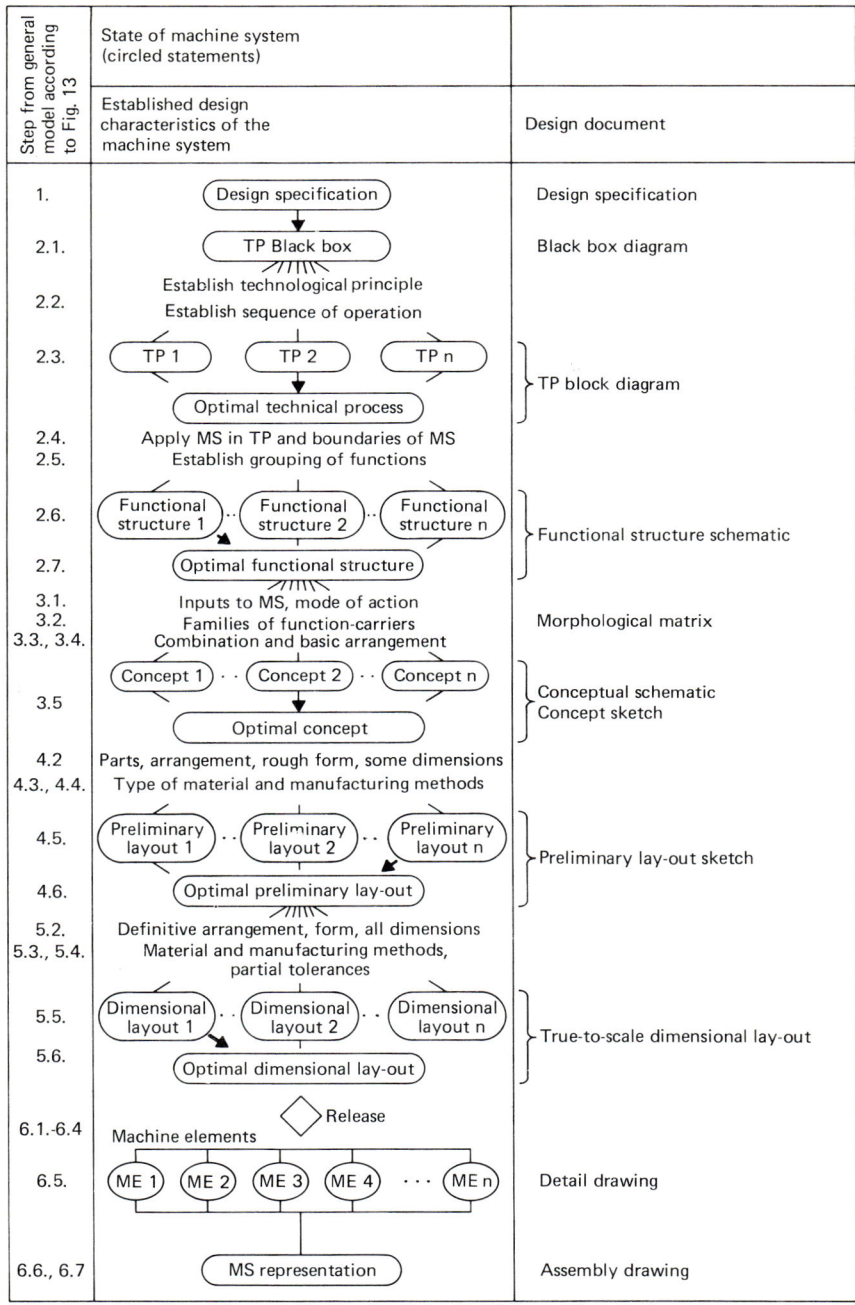

Fig. 14 The Design Process: Graphical Representation of the States of Technical Systems during Design Work.

5.3 EXPLANATIONS OF THE MODEL

To help the reader to understand better the concept of the model, these additional remarks are offered:

(1) *Depth of structuring* in the model is limited to those basic operations for which clear, meaningful statements can be made about the states, characteristics, and features of the technical system to be designed.
(2) A *representation of the procedural model* may be found in the flow diagram, Fig. 13, in which all activities and TS-states are shown, and also in the block diagram. Fig. 14, which shows schematically the design states of a particular technical system. Both representations are complementary. A useful aid to visualising the procedure is provided by Figs 11 and 12.
(3) A division into four stages—problem statement, conceptualisation, layout, and detailing (roman numerals)—is used, even though it performs no useful purpose here. The six phases marked with arabic numerals are to be regarded as basic divisions.
(4) The model is not only usable as a whole, but is applicable in its individual parts and in various sections of the design process. This also creates a link to intuitive procedures. If a partial solution (intermediate state) has been obtained by an intuitive process, one can either review progress using the methodical model, or continue directly with the systematic procedure. In other words, the methodical working mode is not necessarily in conflict with the intuitive mode.

The systematic approach is also intended to provide (pre- and post-hoc) rationalisation and control functions, because many people consider that human minds naturally work in an intuitive mode. For team work, a systematic procedure and record-keeping method is essential to encourage co-ordination and cross-fertilisation between team members, even though much of the individual designer's work may rest on sub-conscious mental processes. The methodical approach provides for a good 'preparation' of the problem, a subsequent 'incubation' period can permit the mind to bring forward its own latent ideas. Post-hoc rationalisation can then provide a link back to systematic procedures. The need for systematic design methods is founded in the inadequacy of intuitive procedures *when applied in isolation*, and in the need to avoid expensive re-thinks during large-scale design tasks performed on high technology systems.

5.4 PARTICULAR STEPS IN THE MODEL

Before we begin describing the separate design steps, a few comments may be useful:

(a) In order to provide easy orientation, the same order of headings is used in each section of the following discussions.
(b) In the given descriptions, no definitions of terminology are given. These are either contained within the previous chapters, or are given in the glossary of terminology, Chapter 9. It is very important for the

reader who wishes to understand the model and explanations to know the precise meaning of each term as used in this book.
(c) It is not easy to select examples which illustrate the general statements. A central case study (Chapter 6) should help to show all the states of a machine system by means of an understandable, fairly simple, design task. This example cannot provide a convincing demonstration of all nuances of the design characteristics. Further examples would be needed.

1. Elaborate or clarify the Assigned Problem Statement into a complete Design Specification

(a) *Entrance*: Setting of the task (assignment of the problem) as a series of requirements or problem situations formulated by a sponsor.
(b) *Exit condition* of the technical system: a complete, quantified and classified set of requirements.
  Design document: List of requirements, the design specification. This is not only the contract document as agreed by the customer, but also contains those requirements generated by conditions within the company, including all statements that could be used as evaluation criteria in later design stages.
(c) *Objective*: provide a complete basis for the design task.
(d) *Defined characteristics* and features of the technical system: usually none, the specification should be neutral with respect to solutions. Only in special cases should design characteristics or features appear as requirements (e.g. in military contracts).
(e) *Suitable methods*: (compare Table 3) market research analysis, checklists, questionnaires.
(f) *Case study:* Tables 5 and 6.
(g) *Comment*: the entry (problem definition) is very variable, particularly with respect to the type of task (whether externally or internally assigned) and other conditions, e.g. the state of the art in that area. The designer is given the requirements for the system to be designed. These are usually minimal, *incomplete* and *unclassified*, and even at times *unrealistic* and *contradictory*. Fig. 11 shows this inconsistency by deficiencies of requirements in many classifications—the radial 'defined requirements' vectors at this level do not cover the '100% fulfilment' range. In this first step, one should recognise the exact problem situation, the nature, scope, and circumstances of the problem, and how the problem situation will behave; one should then attempt to describe in detail what a potential solution to the problem must *do*, without stating what it must *be*. This is aided by a *critical study* of the given task, and analysis of the whole problem situation from all possible viewpoints. Knowledge of the *state of the art* and sciences serves as source of information for comparisons and for *criticism*, and for a rough review of *feasibility*. Market research is assumed to have been performed in the planning phases, but can also be employed here. Review of the technical and economic realisability of a viable solution to the task helps to increase the likelihood of a success and can avoid wrongly applied design effort.

The state of the art contains implicit general knowledge of achievable technology, and detailed knowledge of products on the market. One aim of design can be to achieve and remain at the forefront of this state of the art. The dangers of relying on post hoc market information, thereby working into an established or declining market, but lagging behind the competition, must be weighed against the opportunities and risks facing the market leader.

In addition, one should *classify and order* all requirements, collecting them into classes as shown in Fig. 5, and setting *priorities* (e.g. fixed requirements, minimum requirements and desires). Every requirement should if possible be *quantified* and the *variability* (tolerance) of individual values given, i.e. the categories 'quality, quantity and tolerance' should be stated. This procedure provides the basis for using the elaborated requirements as evaluation criteria in later design stages. Requirements may be differentiated in task-specific or in general ways. A large number of the general requirements are contained in the lists of working principles (Table 4). Viewed in terms of the solution, each requirement is to some extent a *restriction of the solution field*. A relatively incomplete list of requirements permits more solutions than a more comprehensive one.

Care must be taken with the *formal presentation* of the requirements. One should attempt to attain a clear, unambiguous, and solution-neutral formulation. Sometimes a negative requirement—what is *not* allowed to happen—is an advantageous form of specification statement. An easily assimilated and unified form of presentation is also important for the elaborated design specification. It should be accompanied by commentaries and documents relating to negotiations with the contract customer. Organisational preparation and planning also belong to this phase. As shown by the *feedback loop of information* in the model (the dashed lines), the design specification is constantly improved and altered as new questions arise during design work.

2. Establish the Functional Structure

(a) *Entrance*: design specification.
(b) *Exit condition* of the technical system: abstract representation, which projects the TS with respect to its effects (purpose function).
   Design document: functional structure, block diagram.
(c) *Objective*: a largely optimal set of duties for the TS.
(d) *Defined characteristics* and features of the technical system: technological principle, boundaries of the TS, its output effects, and its assembly groupings.
(e) *Suitable methods*: abstraction, black box, technical process.
(f) *Case study*: Figs 18, 20, 21.
(g) *Comments*: It has been mentioned above that it is not possible by discursive means to reach a direct solution of the design task, namely by stepping directly from the requirements to the anatomical structure. More accurately, it is only thinkable by an intuitive leap (which frequently also implies 'jumping to a conclusion'). Normally, one must determine the essential core of the problem, the *output effects* (which in this case are the

purpose functions), that serve as the starting point for the search for an appropriate anatomical structure.

To achieve the output effects we must consider the *technical process*, and this path is marked by five determinations and decisions: these concern respectively (i) the technological principle for the process, (ii) the proportion of involvement of the TS (as compared to involvement of the human) in the process (i.e. degree of mechanisation and automation), (iii) the boundaries of the technical system, (iv) the sequence of operations within the technical process, and finally (v) an optimal functional grouping. The available possibilities for solving these individual design characteristics and features are responsible for achieving *alternative solutions* to the assigned problem (Fig. 14):

(1) Various technical processes arise, depending on the choice of technological principle and sequence of operations.
(2) Various alternative functional structures can arise, depending on the amount of *participation of the TS* in the process, the *formation of boundaries*, and the *grouping of functions*.

When setting out the alternatives, we may *reject* those that can be shown to be not meaningful, or that have severe disadvantages for the given boundary conditions. The selection of the *optimal alternative* implies in practice that decisions must be taken about the optimal design characteristics on the basis of a more or less objective evaluation. As the work on this abstract level approaches completion, one can obtain only a relatively small number of criteria to use in this evaluation; much therefore depends on the experience of the designer.

(h) *Comments to individual steps*:

A. Abstracting (step 2.1)
One abstracts from all the requirements (i.e. one re-defines them in more abstract terms, and where appropriate selects only the most significant meanings), and represents the scope of the problem in a black box model of the technical process (see also Krick (67), Fig. 7.1).

B. Establish the Technological Principle (2.2)
Every transformation of the operand can be accomplished in a variety of ways. The selected technological principle induces some decisions on the basic structuring of the technical process, the work procedures, and their sequence. If for instance the process of milling, of broaching, or of turning is selected for transforming the form (in this case the geometry) of a workpiece, this decision determines which types of work processes (forming operations) are necessary, and usually also their basic sequence.

C. Establish Sequence of Operations: It is also frequently possible to alter the sequence of operations. Taylor's system of work study can be used in this broad fashion for sequencing manufacturing processes.

D. Establish Technical Process (2.3)
The main flow in the technological process is the transformation of the operands according to the selected technological principle, or a combination of such principles (Step B). In order to perform a complete elaboration,

knowledge regarding the *preparation, execution, control* and *completion phases* (TP 4a) is also required. This main flow is supported and augmented by a series of further *material, energy*, and *information flows* (TP 4b). Fig. 2 shows the *complete structure* of the technical process. Experience shows that the arrangement used in this figure should be closely followed; alternatively the design engineer should develop and follow his own appropriate system.

When elaborating the block schematic of a technical process it is possible to establish either the complete structure (i.e. including auxiliary, propulsion, etc. effects), or *only the main flow*. The second alternative seems better, because it provides a clearer view of the prime purpose. The additional effects are subsequently added to the functional structure, as in the model of Fig. 4. The *alternative technical processes* arise from a meaningful application and combination of technologies and sequences of operations. The optimum alternative is then found by an evaluation.

E. Application of Technical Systems and Establishing their Boundaries (2.4)

Each of the effects needed to drive a technical process can be exerted either by a *human being*, or by a *technical system* (TP 1): The *distribution* of these effects between man and the TS can be very variable, depending on which aspects are considered. The degree of mechanisation and automation of the TS is thereby also established (TP 5). These decisions have important socio-political results, especially for project design of large plant. The limitations of human abilities must not be overlooked.

A further consideration concerns the *boundaries of technical systems*. In general, the necessary effects can be realised with one, or with more than one, technical system. If for instance we compare a combine harvester with a functionally equivalent set of separate machines (e.g. reaper, truck, thresher, baler), we obtain an insight into the importance of the decisions relating to this question.

F. Establish Grouping of Functions (2.5)

Considerations similar to those regarding division of work and responsibility that took place in the previous phase are also repeated within the framework of a system. One can group, divide or combine the functions as desired. In the combine harvester example, the various effects require one, or more than one, propulsion device (motor). The consequences of such decisions with respect to various other properties (e.g. economy) can be estimated with some certainty.

G. Establish and Represent Functional Structure (2.6)

The above considerations show that progress towards the *functional structure* takes place by a *meaningful combination of effects* (partial functions) within the boundaries of the technical system. One can select various means of representation, (Figs 20 and 21), but among these only the block schematic fulfils the task of showing interconnections. The model of the functional structure in Fig. 4 indicates a possible form of block diagram that follows from the representation of the technical process of Fig. 1 (see also Fig. 20). The *hierarchical function tree* as shown in Fig. 21 does not fully represent the functional structure, but can be considered as simply de-composing (sub-dividing) the more complicated functions. It is probably preferable as a working basis for establishing the functional structure.

Complementing the main effects (actions) by further *auxiliary effects* has been mentioned under 'Establish Technical Processes' (Step D). The *alternative functional structures* arise by changes of the two design characteristics 'Application and Boundaries of Technical Systems' (Step E), and 'Grouping of Functions' (Step F), by starting from one or more technical processes.

H. Establish Optimal Functional Structure, Improve and Verify (2.7)
The series of functional structures obtained as described above can now be *evaluated*, to establish the *optimum alternative* for the given conditions. Within the usual time limitations, only a minimum of alternatives should be considered. Evaluation is usually based on a *small number of criteria*, because the relatively high level of abstraction of this stage does not permit use of a larger number. The thoughts developed in this phase, and in the evaluation, must be *reviewed* as indicated in Fig. 13, in order to enable a further, more concrete design cycle (99).

Another related procedure encourages the improvement of small deficiencies of a promising alternative. If *weak points* of a solution can be eliminated, the total value of that solution can be increased.

A basic question arises regarding the depth of *detail* to which one can and should sub-divide the total functions in order to formulate the functional structure. In extreme cases, this de-composition can be extended to the *elementary functions* that can no longer be divided. These are defined in various ways by different authors (17, 22, 23, 25), but the general applicability of these concepts for mechanical engineering is not yet proved (35).

From a knowledge of the anatomical structures of existing technical systems, we can learn about the *possibilities for division* of these complicated systems into smaller units of varying levels of complication. If this division is performed according to concepts of *functional units* (partial systems), then an exact correlation between the units and the individual functions in the functional hierarchy is possible. As analysis shows, the division of a given function into partial functions is only possible if the relevant *mode of action* has been previously determined, i.e. one decides to use a particular family of technical system, selected from among a number of possibilities.

It is typical of design work that one works simultaneously at various levels of complication to gradually develop the anatomical structure. A single step from the level of machines directly to the constructional elements, omitting some of the intermediate levels, is generally unthinkable.

3. Establish concepts

(a) *Entrance*: functional structure and design specification.
(b) *Exit condition* of the technical system: abstract, incomplete anatomical structure, formless (shapeless) with the exception of action localities.

Degree of abstraction of the TS: family of technical systems.
Design document: concept sketch.
(c) *Objective*: obtain (conceptualise) a rough anatomical structure that realises the optimal mode of action.
(d) *Defined characteristics* and features of the technical system: sum of inputs, sum of principles and modes of action, therefore also classes of organs as function-carriers, and basic spatial arrangement.
(e) *Suitable methods*: morphological matrix, catalogue of physical effects (a design manual), inversions.
(f) *Case study*: Figs 23 and 24.
(g) *Comments*: conceptual design constitutes the first cycle from the functional structure to the anatomical structure. *In general, this step (the synthesis) is among the most demanding in design work, and in the whole of engineering.* It requires a large amount of imagination, a wide knowledge of available technology from many areas of expertise, a breadth of experience, and the courage to commit oneself to decisions and to lines and words on paper. Criticism after the event is usually much easier than creating something from a blank sheet of paper. The most difficult step, especially for the newcomer, is to draw the first centre-line—the next most difficult step is to rub it out again!

In the spirit of the previous sections, the design engineer must find the *causal action chain of elements* (MS 3) of a proposed technical system, which will deliver the desired effects (as defined in the functional structure) through the effectors to the technical process. The inputs may either be selected by the designer, or are defined in the design specification (Fig. 3). The disturbances from the environment, in space and time, should also be considered. Depending on the degree of complication of the output effects, the action chains and the appropriate function-carriers (organs) will be more or less complicated. With technical systems of second to fourth class (assemblies, machines, and equipment or plant) the anatomical structure must still be sub-divided (decomposed), especially if the proposed effects (partial functions) have been allowed to remain in a more complicated state when sub-dividing the functions in the previous design phase.

The first step in searching for the anatomical structure (action chain) is to determine the *natural phenomena* that are able to cause the desired *effects*. (The situation is understandably easier if the inputs can be freely selected). The *mode of action* is derived from the physical effects by choosing appropriate action localities, their behaviour, and the classes of materials. Combining the individual action chains of the partial functions with the possibilities of partition or aggregation of the function-carriers leads towards the selection of the total anatomical structure for the technical system to be designed.
(h) *Comments to individual steps*:

A. Establish Input and Mode of Action (3.1)
B. Establish Classes of Function-carriers (3.2)
In order to realise the effects in the functional structure that were determined in the previous stage, we pre-suppose the existence of the natural phenomena that could realise such functions. We are therefore searching for *effects* that are usually known as *laws of nature*. Within these

laws, the participating properties are brought into qualitative and quantitative relationships. From the needed effects, one finds the *mode of action* by establishing *action localities* (number and form, e.g. space, surface, line), the *behaviour of the action localities*, and *classes of materials*. It is clear that all the design characteristics mentioned here can lead to *alternative modes of action* by employing different possible embodiments of these design characteristics.

Let us consider as an example a safety clutch for limiting torsional moment in a driven shaft. One class of solution is a slip clutch, in which torque is transmitted by a number of pairs of friction surfaces that slip under overload. The physical effect used is the law of friction:

Friction force $\quad P = \mu N$
Torque $\quad\quad\quad\; M = PRn$

where μ = coefficient of friction, N = normal force, R = radius of friction surface, n = number of friction surface pairs.

The friction surfaces — here the action surfaces — can in principle be selected either as a pair of cylindrical, or flat (end), or conical surfaces. They can be arranged as a single or multiple pairing (e.g. multi-plate clutch). In addition, the materials should be selected such that the friction coefficient is sufficiently large, but that simultaneously the resulting heat can be conducted away from the friction surfaces (e.g. by metal carrier plates, by oil immersion, etc.). Realisation of the normal force N is a further duty for the designer. This problem can be solved using basically mechanical, hydraulic, or electromagnetic principles, e.g. a spring working within Hooke's law, that also permits some adjustment and thereby a reasonably accurate setting of the overload moment.

In this clutch, another effect described as 'transmission of the moment between shaft ends and hubs' must also be realised. If a mechanical principle is selected, the choice is between an assembled friction connection (e.g. a shrink fit as the function-carrier), or a form-locking connection (function-carrier as taper key, parallel key, transverse or longitudinal pin, spline, serration, etc.; note that there can be some overlap between friction- and form-locking principles, some embodiments can be regarded as belonging to either class). These examples demonstrate a few of the classes of function-carriers that can achieve the necessary effects.

An advantageous *method of recording* this information to prepare for design decisions of this nature is the *morphological matrix* (Fig. 23) (65, 98). The first column lists the partial functions extracted from the established functional structure, and each matrix line is used to record the *functional principles able to realise* these functions, and corresponding *classes of function-carriers* (organs). A basic question in this context is directed at the *number* of modes of action and of function-carriers that should be generated and recorded; one could conceivably cover all possible variations of these items. It has been found best at this stage to *reject* solutions that represent unimportant variants, and to realistically restrict the solution field according to the boundary conditions, and the nature of the assigned problem.

This conclusion is valid at the state of detail and level of abstraction shown here. At a higher level of abstraction, when searching for possible novel solutions which could represent potential new product developments, concentration on a far greater range of principles is advisable, but details of function-carrier should be restricted. There are two opposing dangers: one of artificially restricting the solution field by 'mental set' (fixation, prejudice against some possibly viable solutions); and the other of generating far too many solutions and thereby losing the essential overview.

C. Combine Function-carriers and Examine Relationships (3.3)
A total concept is theoretically generated by forming every *combination of all appropriate function-carriers* to form different anatomical structures. Proceeding in this fashion, the number of solutions would be far too large; one soon discovers that not all solutions are meaningful, that not all elements are compatible, fit together, or work together. It is therefore recommended that the engineering designer should concentrate on meaningful and realistic combinations, in order that the subsequent effort expended on evaluation is not too large. (This does not mean that solutions *outside one's immediate experience* should be rejected; such solutions may be closer to the state of the art, and can even provide a competitive edge over other manufacturers and suppliers. Learning is, after all, a life-long experience, and stagnation can lead to obsolescence and demise.)

Finally, the resulting combinations should be carefully reviewed, not only for the effects and functional abilities, but also for compatibility of elements, and *fulfilment of other properties*, particularly the functional parameters such as power, speed, size, etc.

D. Establish Basic Arrangement (3.4)
At this stage, the elements of the anatomical structure as established exist only in a *functional relationship*. Using the example of the torque-limiting slip clutch, the established mode of action is 'that a spring presses two friction surfaces (of a pair) together, one of which is driving, and the other driven'. One does not at this stage need to make any decisions about the types and numbers of springs, about their positions, their dimensions or materials. The *representation of the concept* is intended to show a rough idea of possible *arrangements* of function-carriers, as sketches of the 'relative spatial relationships among the constructional elements'.

Such a conceptual representation should prevent the fixing of a poor idea for the next processing step. It is therefore necessary to present a well thought out rough arrangement. One must consider the basic possibilities of arrangements, i.e. the mutual positions in space of possible constructional elements, and select one or more of the advantageous alternatives as a basis for conceptualising, i.e. forming mental and graphical images of the desired physical reality, the TS. In this case, too, we are concerned about generating *classes of arrangements*, but not at this stage with generating variations within a class.

E. Represent Concept (3.5)
From the above discussion we may conclude (as is confirmed by practice) that the level of concretisation of the concept can range within wide limits, and this is true of all other origination stages of technical systems. This

variability is also reflected in the form of representation which is used to show the concept, it can either depict the abstract anatomical structure as a block schematic (similar to the functional structure) with circles to denote elements and connecting lines to denote relationships (as in Fig. 3), or as a more concrete concept (e.g. a skeleton diagram) as shown for the welding positioner in Fig. 24. Even within such representations one may find many different stages of concretisation.

With some exceptions, the representation of concepts in machine systems is not standardised, therefore their interpretation can be very varied. There are some areas, e.g. hydraulics, electrics, thermal technology, in which a relatively small number of classes of constructional elements exist, and these are designated by agreed symbols, such as those for valves, switches, transistors, and similar items. This excursion away from the purely mechanical into other fields permits us to discover other noteworthy and explanatory analogies. For example, a hydraulic or electrical schematic, in which data for the elements is not recorded, is classed as a general concept for that system; a particular system with certain predicted capabilities (design properties) is generated by choosing particular elements to create a definite concept. If two different functional principles are recognised as being capable of delivering analogous functions and effects, such a system may be realised with either of the two principles, and therefore employing the appropriate constructional elements (as with electronic and pneumatic logic control systems).

F. Establish Optimal Concept, Improve, Verify (3.5)
The concluding phase of conceptual design again contains the procedures of a *main evaluation*, an *improvement of weak positions*, and a verification. In addition to the consideration existings from Section 5.5, some comments are needed to the situation at this stage. As the statements describing the technical system are still very abstract, the overall evaluation is difficult. The established design characteristics give too little leverage for quantifying most of the requirements that may be selected as evaluation criteria. Particularly the economic area suffers a distinct lack of accurate evaluative statements.

4. Establish preliminary layout

(a) *Entrance*: conceptual sketch, functional structure, and design specification.
(b) *Exit condition* of the technical system: roughly dimensional description of the anatomical structure of the TS.
  Degree of abstraction of the TS: type, genus.
  Design document: layout sketch.
(c) *Objective*: first rough iconic representation of the structure with some important functional dimensions (roughly to scale).
(d) *Defined characteristics* and features of the technical system: constructional elements, their basic form, important dimensions and classes of manufacturing methods for the main parts, and basic arrangement of the constructional elements.

(e) *Suitable methods*: variation of characteristics, inversions, value analysis (33, 34, 57, 63, 79).
(f) *Case study*: Fig. 25.
(g) *Comments*: The path from concept to dimensional layout takes two steps in our model. In the first one, a preliminary layout is produced to describe the rough anatomical structure. Based on starting dimensions that are as reliable as possible, loading histories, and a rough anatomical arrangement as produced in the previous phase, it is possible to begin form determination of the action localities. As one is working with a large number of assumptions in this part of the form-design phase, it is preferable not to solve all the details, but rather as shown in our model to create a series of rough layouts that represent possible technical solutions with respect to the most important requirements.
(h) *Comments to individual steps*:

A. Establish Orientation Points for Form Determination (4.1)
As starting point for establishing the form of the constructional elements, some *orientation points* (e.g. starting dimensions) should be established. This is aided by making comparisons with similar technical systems or functions, and by *approximate strength calculations* based on reasonable assumptions, because accurate data of loading cannot be determined. Above all at this stage, wide experience, a good collection of information, and suitable working means (such as data sheets, nomograms, computer programs, etc.) lead to quick and reliable results. Such calculations are not only needed at the start of layout work, but continue throughout the form-design phase. These calculations again demand good knowledge, in this instance especially about the availability of specialist design information, and also a wide knowledge and experience of possible modes of failure of components and mechanisms, and of the factors that influence failure probabilities. Reliability can be made or marred by the skill and capability of the design engineer in this concretisation phase.

In advance of such calculations, a basic qualitative study of loading conditions should be made, so that the general form respects the load transmission paths. The ideal of uniform stress distribution is usually not the aim, but rather a sensible, well proportioned, strength-related preliminary form (18, 28, 75).

B. Arranging, Re-use, Form-giving and partial Dimensioning (4.2)
Starting from the concept, one can undertake the accurate *arrangement* of partial systems. There are many possibilities for alternative arrangements; the solution of this problem is central to producing the rough layout. The second major task is then establishing the *basic shape* of individual parts. The problem is simple if an existing part can be *re-used*, e.g. as standard part, bought-out component, or repeat part. That part can immediately be drawn in, with its correct form and dimensions, and can often be used as a guide to initial size for determining a *proportional form* for adjacent parts (for example, sizing some of the coupling parts to suit the shaft diameter for an electric motor to which it could be fitted). The economic consequences of re-use of components are generally well known.

C. Material Types, Classes of Manufacturing Methods, Tolerances, Surface Properties Established for Individual Cases (4.3)

In general it is not necessary to determine these design characteristics in the rough layout. The exception is that of the action localities, where a material specification should be entered because it is an already determined property. Similar treatment should be given to other design properties: as soon as they are fixed, they should be entered on the sketches. This is also valid for bought-out parts that should be labelled with correct catalogue numbers and leading dimensions.

Another reason for considering these problems lies in the process of form determination itself. Just as the form depends on the choices of materials and of manufacturing or assembly methods, so also does the arrangement of parts. Important information for the subsequent design steps can be recorded on the representation: much of the information is contained directly in the form.

D. Investigate Critical Form Determination Zones (4.4)

In every technical system one finds regions, termed *critical form determination zones*, that play an important role in choosing the total solution from various aspects, or that exert a large influence on the possibilities of creating variants, and thereby of optimisation. These critical form determination zones should be given particular attention: if needed, they should be graphically represented on an enlarged scale, and a number of alternatives worked out.

E. Represent Preliminary Layout (4.5)

Representation of the rough anatomical structure is best performed in a *layout sketch*. This should be a clean free-hand pencil sketch, roughly proportional and to scale, and stating some typical dimensions. Its projection can vary, either orthographic or axonometric (e.g. isometric, trimetric, oblique, perspective, etc.). As there are no definitive standards for this type of representation, key-word notes, explanations, etc. may be used on these sketches.

F. Establish Optimal Preliminary Layout, Evaluate, Improve and Verify (4.6)

Evaluation of alternative anatomical structures recorded in the layout sketches is more accurate, because many of the design characteristics are already fixed; from them the values of various selected criteria may be deduced. Because it is now possible to determine or estimate data that influence manufacturing costs, (e.g. weight, size, basic metal cutting data, etc.), this is the first opportunity to perform a *rough cost calculation*. The evaluation should indicate the weak points of individual alternatives from various viewpoints, and permit improvements where desired.

5. Establish dimensional layout

(a) *Entrance*: layout sketch and all other documents to this point.
(b) *Exit condition* of the technical system: complete, almost finalised description (graphical, verbal, and numerical) of the anatomical structure,

dimensions at least recognisable in a true-to-scale representation.
  Degree of concretisation of the technical system: species or serial sizes (refer Fig. 8).
  Design document: dimensional design layout.
(c) *Objective*: clear, complete anatomical structure with all constructional elements and their arrangement determined, production of a co-ordination document for detailing.
(d) *Defined characteristics* and features of the technical system: fairly well established form of the constructional elements and their arrangement, most dimensions, materials, manufacturing methods, important tolerances and surface properties.
(e) *Suitable methods*: strength calculations, value analysis (33, 34, 57, 79), systematic form determination (18, 28, 75).
(f) *Case study*: Fig. 26.
(g) *Comments*: In the second cycle of layout work, we are dealing with clarification and further concretisation of the selected preliminary lay-outs. During this stage, the arrangement and the elements should be finalised. The main concentration should be on definitive form determination, dimensioning, and material selection, based on strength calculations or strength verification for important parts. If a new idea emerges, we should not exclude the possibility of creating a revised dimensional layout that differs appreciably from the preliminary layout.

As the individual activities are almost repeats of those for the preliminary layout stage (only with a change of emphasis), it does not seem necessary to give extensive comments to most of the partial steps; only to those that need further amplification are mentioned.

A. Delivery of Substantiation for certain Design Characteristics (5.1)
For most designers, this step is completed by performing the strength verification (e.g. 'stressing') (26, 51, 70, 86, 88); the number of loading cases to be substantiated in a particular problem is subject to judgement. The problem is, however, somewhat broader, the other extreme being complete testing and experimental substantiation of most dimensions, form, materials, manufacturing methods, tolerances (44, 71, 82), and surface properties. Establishing why a particular dimension, form, etc., has been chosen, and not another, is a major challenge for the engineering designer. This is especially true with respect to the ergonomic (28, 66, 90, 97) and aesthetic (40, 76) properties. A close connection with the operation 'evaluation' is emphasised here. The execution of the calculations, considerations, and proofs is distributed over the whole of phase 5, and finishes with recording (or filing) the results in the *technical report*. This key-word indicates another problem area, the significance of which can only be hinted at.

E. Representation of Design Layout (5.5)
Only a few comments are offered with respect to the dimensional layout. The representation of the complete anatomical structure should be performed true to scale, and using full drawing equipment (at least a straight edge and compasses). These lay-outs are mostly sectional drawings in which the individual parts can be represented. Dimensioning is extremely variable, from an almost complete lack of dimensions to almost fully

dimensioned. A judgement of advantages or disadvantages is hardly possible, because of the many factors involved. The quality of design engineer or draughtsman (and of manufacturing personnel, in some one-off situations) can determine which level is appropriate for recording this information.

F. Establish Optimal Dimensional Layout, Evaluate, Improve, and Verify (5.6)

Both *optimisation* (64, 78) and *evaluation* are of greatest importance at this stage, not only for the quality of the technical system, but also for the economics of the design process itself. Even if only a small proportion of the total costs have been used in reaching this stage, the costs will escalate rapidly during detailing.

The *evaluation of the layout* should now be performed with best possible accuracy, depending on the amount of time that we wish to invest. It is important to decide whether this evaluation and the relevant calculations are to be performed only by the design engineer, or whether specialists should be called in. In any case, most of the design characteristics are decided, and the rest can be estimated; therefore the basis and data for all calculations should exist.

The importance of evaluations and decisions at this stage is emphasised by the step labelled *release for detailing* indicated in the model (99). It is usual to employ a review commission or syndicate, delegated from various areas of expertise or specialisation. Value engineering and cost-benefit analysis (68) can usefully be applied as review techniques. An evaluation can now be very objective, because the degree of concretisation attained at this stage allows all participants on the syndicate to obtain a fairly realistic impression of the technical system, much better than at the conceptual stage.

6. Detailing, elaboration

(a) *Entrance*: dimensional layout.
(b) *Exit condition* of the technical system: complete description of the anatomical structure and all constructional elements with all necessary data.
   Degree of concretisation of the TS: type or particular model size.
   Design document: workshop drawings, possibly numerical control tapes for manufacturing, if CAD-CAM techniques are used.
(c) *Objectives*: finalised and complete description for manufacture.
(d) *Defined characteristics* and features of the technical system: all design characteristics.
(e) *Suitable methods*: none specific.
(f) *Case study*: no illustrations.
(g) *Comments*: This last phase is terminated by the complete description in drawings, reports, etc. of the proposed technical system. The dimensional layout that served as entry point defined some of the design characteristics completely, but others were not ever considered. Accordingly the scope of work in detailing changes with the amount of information presented in the layout. A number of *calculations* such as the remaining strength verifications, weights, geometries (e.g. for gears) support the partial

decisions. Above all the aspects of manufacturing properties, and the economics of manufacture connected with them constitute the centre of concentration.

In fact, a large proportion of the design cost is incurred in this phase, and the small detail decisions account for a very large proportion of the manufacturing costs of the product. Skills, knowledge and ability are essential, especially at the detail design level, because a very minor change of geometry, or omission of a radius and transition of surfaces, can lead to early failure (e.g. by fatigue) or can unnecessarily increase manufacturing costs (e.g. by demanding special processes or tooling). Even (or especially) at this level of concretisation, the problems are worthy of the attention of professional (chartered) engineers.

A significant proportion of the time spent on this stage is necessary for the *workshop-true representation*. The impression that an observer obtains from the time and effort expended at this stage on apparently routine activities is frequently regarded as the characteristic of design engineering. This idea is not correct, because even if no further conceptual decisions are made during detailing, the design time can be used to great advantage to verify previous decisions, and to perfect some of the properties.

Checking of drawings

Checking of drawings is emphasised as a separate step of great importance. Each drawing should be examined more than once from each viewpoint, so that preferably no queries or alterations occur during manufacture. We will refrain from making additional comments, because this phase consists of known activities that are fully described in other areas of knowledge.

5.5 GENERALISED BASIC OPERATIONS IN THE DESIGN PROCESS

The previous section dealt with the individual phases and steps in design engineering, as they take place in solving a problem. This sequence consists basically of operations from the first and second levels of Fig. 10. Their objectives are always to find, improve, select, etc., some of the properties of the technical system to be designed.

In all these activities there is a group of frequently repeated basic operations that appear in the flow diagram (Fig. 13), but have not so far been discussed in detail. Mostly they are implicitly contained in the whole procedure, and also in individual steps. The four basic operations—definition of problem (elaboration of assigned specification), search for solutions, decisions, and communication of solution—are an invariable sequence, supported by the results of the subsidiary operations of providing and preparing information, verifying, and representing. This cycle of four operations forms a repetitive set of activities, usually starting either from the search for solutions, or from the definition of the problem.

5.5.1 Elaborating the assigned problem statement into a specification

At the start of the design process, the problem statement for the total system was discussed. This is only *one of many problem statements* that must be elaborated into a working specification for the current stage of the design process. Such Design Specification statements must be worked out, consciously or unconsciously, for all partial systems and elements, as well as for the total system, and in each design step. The importance of the problem statement can be seen from the saying: a correct problem specification is half a solution.

The problem statement declares the *objective of each activity*; if it is not clear, then the result cannot be optimal, nor can the path to a solution be rational and quick. The following aims and principles are a guide to the elaboration of the problem into design requirements (specifications):

(1) The design specification should be *complete, orderly, clear and unambiguous*, anticipating reasonable progress compared to the state of the art.
(2) The design problem and requirements should be *formulated by stating effects* (what the system should do), not the means of accomplishing them (what the system should be), e.g. 'a shut-off and regulation organ for liquid flow', not a valve (65).
(3) Requirements should be *classified* as fixed, minimum or desired.
(4) Requirements should be *quantified* and toleranced where possible, e.g. temperature 200°C, but also ± 10°C.
(5) The problem statement is a set of information, as is also the list of requirements; therefore knowledge from the field of information processing is applicable (e.g. establish the state of the art). In this stage, the majority of the technical information concerning the problem and solutions should be collected. The term *study stage* is frequently applied to this process, but it would be wrong to assume that this concludes the activity of information gathering, it runs through the whole design process.

In the framework of the study stage, the state of the art is established as shown in step 1.2 of Fig. 13. This phase also provides an opportunity for discussing the problem situation (Step 1.3), and particularly to obtain knowledge of the *causes*. In an extreme case, one can even find a solution here. Widely variable problem statements are possible, according to whether the requirements are set as (a) a problem, or a deficiency, or (b) a need and a state of the defined object, or (c) as a necessary effect (functional requirement), or (d) as a very definite technical system (requirements and technical means). Appropriate considerations are discussed in Chapters 2 and 3 (TP 1 and 2).

An examination of the *possibility of realisation* (feasibility study) of the system at this stage (Step 1.4) should show whether the problem can be solved at all, and whether it is useful to solve it. Feasibility in this context is generally regarded as a measure of the *probability of realisation* of the system, estimated in advance. As probabilities can never reach the value 1.0 (unity, or 100%), thus certainty is never possible. Even if a system is fully realised (showing certainty that it functions correctly), it only has this 'certainty' level for a shorter or longer period of time, i.e. until a failure occurs. Thus, feasibility can be *disproved* at this stage, but feasibility can never really be proved. A strong indication (high probability) of potential success is all that can be hoped for, but this constitutes a favourable result.

In this feasibility study process, the following areas should be investigated:

(a) *Technical*: is conformity to the laws of nature possible? (Does the proposed system, solution, or requirement conform to the natural laws? Is a perpetuum mobile demanded?) Are enough technical means and experience available to facilitate at least one solution?
(b) *Economic*: can the operation of the machine system and its manufacture be economic?
(c) *Financial*: is the situation favourable, and are sufficient financial means available to solve the problem?

An important section of the work to be performed in developing the problem statement is *planning and organisational preparation* for the continuing activities. An impression of the likely progress of the solution process in time and space, and of the necessary means (financial and material), are among the essential and purposeful prerequisites. This area is closely related to the management of the design process (47, 48, 69, 95).

The working methods employed should contribute towards reaching an appropriate level of completeness and quality. Of the known methods, the most suitable are (compare Tables 3 and 4):
• Brainstorming: new ideas.
• Attribute analysis: new ideas, quality, quantity.
• Method of systematic field coverage: completeness.
• Method of questioning: completeness, quantity.
• Market research analysis: quality of requirements concerning the market.
• Network planning: overview of timing and sequence (e.g. critical path technique, or PERT).
• Method of systematic doubt: examination of all given data, scientific scepticism.

The phase of elaborating the assigned problem is particularly suitable for *rationalising the work*. One of the important foundations is a survey of all the properties that are relevant to that branch of engineering; the starting point is, for instance, given by the categories of properties of machine systems as shown in Fig. 5. A practical example of use is shown in Tables 5 and 6. It is of advantage to collect functional parameters over a period of time in the form of survey tables and graphs, to discover trends of development. This question has been discussed in Section 3.7, 'Development of Technical Systems'.

It must also be mentioned that the *character and personality of the design engineer* plays an important role in determining the quality of the problem statement. Because the quality of the requirements is strongly dependent on the design engineer, particularly in situations where he establishes these for himself (usually for all partial and elementary machine systems), the project group leader's (and the designer's) awareness of responsibility is critical for the quality of the specification.

Agreement on the clarified problem statement between the design office (as representative of the manufacturer) and the customer (whether a real one, or a potential one represented by the sales office) takes place formally in the *technical requirements* that should form an important part of the contract. The contract documents contain agreements about all relevant properties of the future machine system, and also conditions for delivery, commissioning, determination of the guaranteed parameters, etc. Pre-printed proformae (e.g. see 69) can assist in this process. The prepared and accurately formulated requirements on the technical system are often termed the *design specification* (list of requirements, duty booklet).

There are great differences in practice in defining project requirements. At one extreme, design of large plant, a project team including engineers can negotiate the contract with the customer, and the terms and conditions of the contract may be formally and legally sealed. At the other extreme, the sales office can dictate requirements to the design office—a tendency to be resisted, because the partnership may be one-sided. This can lead to real difficulties in interpretation of requirements, and of motivation for the design engineer. The sales office must, at most times, play the role of intermediary, but should not be permitted to eliminate the design office from the negotiations, even when the sales staff includes qualified professional engineers.

5.5.2 Searching for solutions

The second basic operation is concerned with searching for solutions, i.e. for *physical means* that eliminate the deficiencies, or that achieve the desired conditions of the operands, or that attain the desired effects and other properties. For the design engineer, the centre of attention is particularly the output effects that he should obtain with his proposed technical system. As has been shown above, he wishes primarily to determine the action localities and design characteristics, in order to define the desired anatomical structure. We have described two paths towards finding solutions:

(1) The *intuitive*, when in a single leap the totality of a solution emerges via unconscious or preconscious thought.
(2) The *methodical, discursive*, in which the optimal solution is approached in small conscious steps, usually iteratively.

The methodical way is based particularly on the following principles:

(a) *Division* of a complex problem into sub-problems.
(b) Progression from *abstract to concrete*.
(c) Individual design characteristics (that give rise to variations) are *progressively* fixed after careful optimisation.

Because the search for solutions is the peak of creative work, many theoretical investigations have dealt with this operation. As a consequence, a large number of methods have been proposed to assist in problem-solving, i.e. for searching for appropriate means (machine systems in the case of design problems). Of the methods shown in Table 3, those *suitable for the search for solutions* are:

Classification of Design Manuals (35)

Solutions Manual (Force Amplification) (Part from Roth (24))

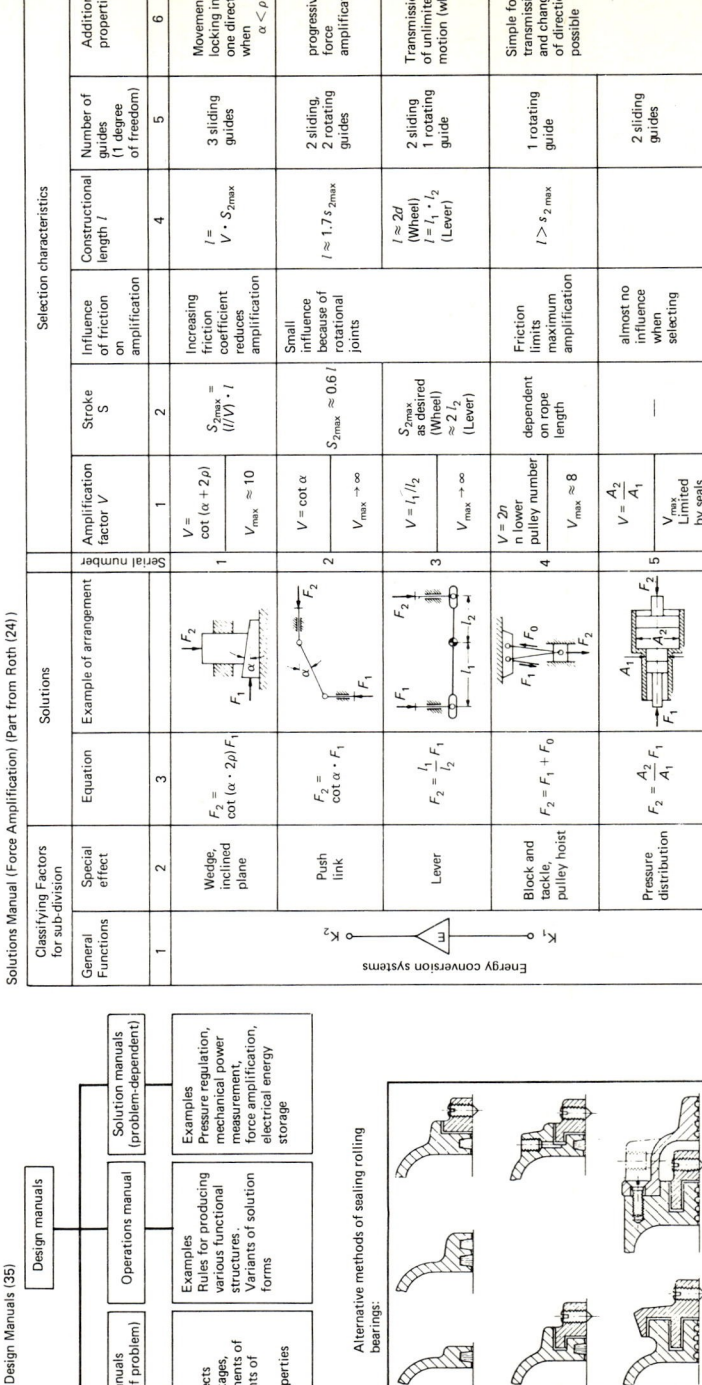

Alternative methods of sealing rolling bearings:

Solution Manual (Decision Table) (Transmission of Torque)

Parameter Matrix for Function; Transmission of Torque from one Shaft to Another

Function-carrier Families of TS that fulfil the function	Important parameters or characteristics						
	Possible shaft orientation	Maximum ratio i_{max}	Maximum power P_{max} kW	Maximum circumferential velocity v_{max} m/s	Average efficiency η	Average life Operational hours t_L	Guide to relative cost	etc.
Straight-tooth spur gears	‖	4 (7)	$50 \cdot 10^3$	15 (100)	0.98	10^5		
Helical spur gears and angled drive helical gears	‖ ·∣	7 (10)	$50 \cdot 10^3$	25 (150)	0.97	10^5	Typical values are found for specific cases	
Worm gears	·∣	4 (100)	50 (200)	15	0.4–0.7	10^3		
Friction wheels	‖	7 (15)	100	15–20	0.96	$2 \cdot 10^3$		
Chain drives	‖	8 (10)	100 (3500)	40	0.9–0.97	2–$10 \cdot 10^3$		
Flat belt drives	‖ (·∣)	5 (20)	100 (3500)	35 (100)	0.88–0.95	very variable		
Vee belt drives	‖ (∕∕)	7 (10)	100 (1500)	60	0.96	very variable		
Rope drives	‖ (∣∣)	5						
Tooth belt (timing belt) drives	‖							

Solutions Manual: Transformation of Energy of Signal Type

Cause ▭ Capacitance

Cause:	Physical Effect		Law	Bibliography	Example of Application
01.11 Length, cross-section volume	Electrical condenser (capacitor) (plate spacing)		$\Delta C = C \dfrac{\Delta d}{d}$	[22], p. 150	Length measurement
	Capacitor (surface area)		$\Delta C = C \dfrac{\Delta A}{A}$	[22], p. 154	Variable capacitor
	Movement of the dielectric		$\Delta C = C \dfrac{\Delta A}{A}(\epsilon_1 - \epsilon_2)$	[22], p. 154	Punched card/tape reader
	Thickness of dielectric		$\Delta C = C \dfrac{d_2(\epsilon_2 - \epsilon_1)}{d\epsilon_2 - d_2(\epsilon_2 - \epsilon_1)}$	[22], p. 157	Thickness gauging
Force, pressure, mechanical energy	Plate spacing		$\Delta C = \dfrac{2}{U^2} F \Delta s$ $\dfrac{\Delta C}{C}$	[40], p. 108	Force measurement

Fig. 15 Design Manuals (Catalogues): Survey and Examples.

- Discursive methods: consideration of analogies, aggregation, similarity laws, structuring, inversion method.
- Intuitive methods: brainstorming, synectics, method '6-3-5'.

Selection of the most suitable method is guided by the *complication, type and originality of the problem*, and by further properties of the system to be achieved. Generalising from experiences in this area, a few principles can be defined:

(1) Always find *a few meaningful solutions*, in order to have the opportunity of selecting between them.
(2) *Concretisation too early* can confine the considerations in a particular direction (prejudice, mental set, fixation).
(3) Adopt *existing solutions* in sensible proportions: not every step in the design process can result in an invention, and invention usually carries a higher risk.
(4) *Co-operation* with other experts can bring good results in finding the optimal solution.
(5) *Systematic procedure* is recommended in the step of searching for solutions, but is particularly important in all steps of review and revision.
(6) All solutions should be briefly *noted down (recorded), with comments* where appropriate.

Suitable aids can facilitate the search for solutions, and lead to better results; one such aid is, for instance, the range of surveys of solutions for various functions, i.e. certain types of *design manuals* as shown in Fig. 15 (3, 17, 24, 25, 92, but see also 53, 56) (in the literature, these are also referred to as 'design catalogues').

5.5.3 Evaluating and deciding

The last step in the cycle of basic operations is that of evaluating and deciding. Alternative solutions that have been produced in the previous operations should now be evaluated, and the optimal solution for the given conditions selected. Based on the criteria selected from the list of requirements, one evaluates the solutions as objectively as possible, and declares the one with the highest rating as optimal (16, 65, 75).

The procedure recommended here assumes that the parameters are largely linear in their behaviour. (If this is not the case, then a collection of sub-optima such as would result from this process is probably not an overall optimum. This danger should be kept in mind when selecting the individual optimum partial solutions.) It is usually possible to depart from the 'exact' optimum by a larger or smaller amount, depending on the sensitivity of the properties to such changes. One should also keep in mind that an evaluation by point rating, or weighted point rating, is to some extent subjective. Any differences between the highest and next ranked solutions may therefore not have statistical significance, i.e. a small number of solutions (rather than just one) may have to be carried forward to a more concrete level before a firm decision can be taken.

The *aim of an evaluation* is to be able to make a statement about the quality or value of a system (its total value), or of only one of its properties. A number of useful considerations to this problem may be found in value theory, others in decision theory. Analysis shows that for evaluation during design work the following questions should be posed:

(1) How good is it? (Demands a total or a partial evaluation.)
(2) Does the solution conform to the problem statement?
(3) Which solution is optimal?
(4) Which are the optimal values of some characteristics?

The procedures for various *types of evaluation* are shown as flow charts in Fig. 16; these contain the repeated operations:

- selection of criteria,
- choice of scales for the criteria,
- determination of values for the criteria,
- processing of individual value into a total value.

(a) Selection of criteria for the evaluation

The *quality* (total or partial value) of an object is in general a *vector of the values of its properties*. Besides the total value, one can also form partial values such as technical, usage, consumer-economic, or status (attraction or esteem) values. The evaluation can be based on simple or complex characteristics that are meaningful in relation to the object, and are mostly of an economic nature, e.g. economy, efficiency, profitability, etc. The *selection of criteria* is not a simple matter. One tries on the one hand to characterise the object in as wide a range and from as many viewpoints as possible, but such a combination of many different types of values leads to relatively problematical results that are difficult to interpret; on the other hand the characterisation must sometimes occur with as few values as possible to provide a quick assessment.

A different problem arises with respect to the ability to *determine* values for some of the criteria. During design we are still in a very speculative and abstract region, where determination of criterion values can rarely be objective and single-valued. It would then be meaningless to select those properties as criteria whose values cannot be determined at the current design stage.

(b) Choice of scales for the criteria

The evaluation statement should characterise the suitability of the object for the given requirements; it can therefore vary from 'very suitable' to 'unsuitable'. In technical practice, a four-point quality scale (rating) has been found useful:

very well suited	4 points
well suited	3 points
satisfactory	2 points
barely adequate	1 point

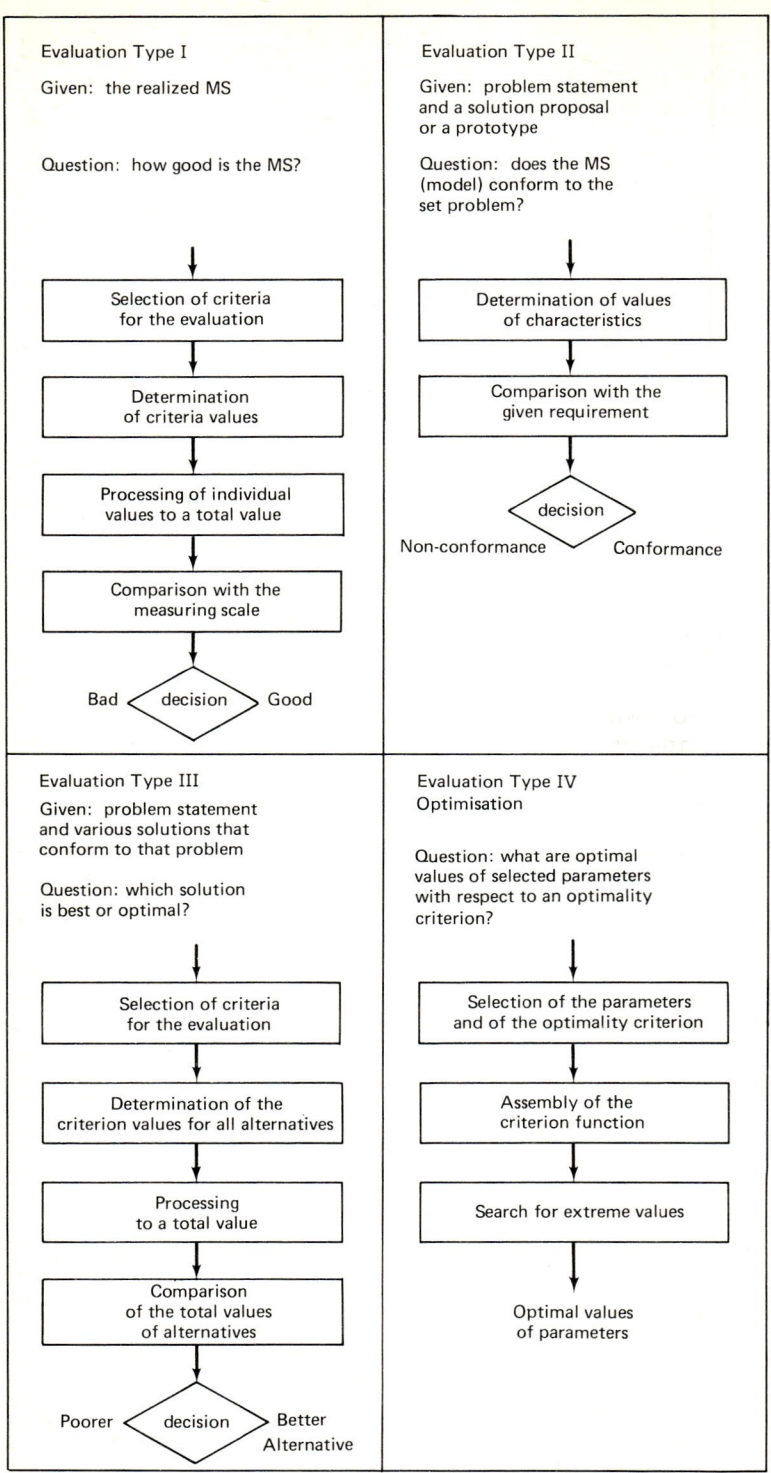

Fig. 16 Evaluation: Procedural Model of some Types of Evaluation.

(A zero rating could be used to characterise a completely unsuitable solution, and this would imply that the solution should be eliminated from the evaluation and from any further consideration, if this rating is attached to an important characteristic.) Ten-point evaluation scales have also been used, but there is a large possibility of error in selecting the most appropriate score (number of points) for many of the properties. A further refinement can be introduced by *weighting the individual criteria*. There is no doubt that the properties of an object have differing importance. The most important can, for instance, be given a ten-point (or four-point) weight, those of lesser importance appropriately lower. The total score for a solution is then the sum of individual products of quality score and weight allocated to each criterion (but see the earlier comments in the second paragraph after the heading to Section 5.5.3 above). An ideal score can also be calculated, by multiplying the sum of weights by the maximum possible quality score.

In order to evaluate certain objects for quality, the selected quality score should correspond to the achieved property values. This is relatively simple for *quantifiable properties*, e.g. height 2.1 m = 4 points, 2.5 m = 1 point. A satisfactory objectivity can also be attained for clearly defined properties, e.g. colour white = 4 points, black = 1 point. There are a number of properties for which an *accurate assessment is not possible* in relation to an objectively defined point scale. These are properties such as safety, or ease of operation. The evaluation must then be based on as objective an investigation as possible. In practice, adequate justifications and reasons must be given for the results of such evaluations.

For some of the evaluations taking place within the design process, a clearer decision may be made on the basis of a two-dimensional 'relative strength' diagram, as proposed by Kesselring (16). The ordinate (vertical axis) shows the weighted-rating sum as a fraction of the maximum theoretically attainable sum (the ideal score): a relative quality of the solution's achievement. The abscissa (horizontal axis) plots the economic ratio: anticipated costs, especially for manufacture, compared to anticipated returns, i.e. a benefit/cost ratio. Any solution scoring greater than 0.7 on both scales is likely to be acceptable. Plotting all available solutions on such a chart can give a clearer picture of the qualities of individual solutions, their capacity or need for improvement, and other factors.

(c) Determinations of values for the criteria

During *experiments* many values of technical systems can be measured, and these then facilitate a clear evaluation statement. The situation is more problematical during design, when it is only possible to use mental models and thought experiments to determine many not directly representable properties (only design properties are directly representable). In such cases, one meets a wide variety of opinions to the same problem. The designer's experience and objectivity are the important factors here. Only after the system has been fully realised can the truth become uniquely clear. Even then, the full truth of objective quality and performance is only clear within the embodiment of the machine system. Human opinion of it can still vary by a large amount. Consider the problem of assessing reliability of a specific new car in the sales room of a dealer. One can only estimate this by considering the reputations of the car's makers and of the dealer. It would take a long series of well-recorded experiments to obtain a reasonably objective value for reliability of the car type, but that assessment is not applicable to the individual car—it may be one of the small rogue population that goes wrong every other day, or it may outlast all others of the same type.

(d) Processing of the individual values into a total value

When the individual values for the selected criteria are available, e.g. as point-scale scores, the problem of combining these to calculate an overall value remains. Practice and theory provide many models for such procedures, from simple addition to very complicated processing methods. It has been shown that not all of these methods deliver reliable results. Details may be found in the relevant literature (8, 15, 19, 22, 33, 75).

5.5.4 Providing and preparing information

Having the correct information at the right time is at present one of the most discussed problems in all branches of engineering. Today's situation is practically the same in design, in research, and in management, and can be characterised as follows:

(1) Over-abundance of information and of information media.
(2) Only a part of the information is usable.
(3) Short currency life of some information.
(4) Not enough unification of terminology.
(5) Abstracts and summaries of documents do not represent the contents well enough.
(6) Too little time is available to absorb information.
(7) Obtaining and processing information is costly.
(8) Knowledge of the information user about documentation systems is unsatisfactory, and this results in low utilisation of available documentation.

We do not wish to discuss the range of problems here, but simply state some important aspects of working methods in this area.

(a) Properties of technical information

We interpret the term 'technical information' (including specialist design information) to mean all recorded knowledge and insights, and remembered understanding and group intelligence that are necessary for design work, from the laws of natural sciences to the data concerning production operations of certain technical systems. The quality of information depends on the following properties:

(1) Correctness of contents
(2) Intelligibility
(3) Uniqueness, clarity, accuracy
(4) Availability at required time
(5) Verifiability (source references)
(6) Completeness (no sections out of context)
(7) Form (surveyable, easy orientation and retrieval)
(8) Type of information carrier or medium
(9) Author of the information
(10) Reachability, ease of access to information
(11) Potential user group at whom the information is aimed

(b) State of the information

The state of information is decisive for its applicability and the possibilities of use. One can distinguish:

(a) Un-fixed information (in human memory), hardly accessible, and
(b) fixed information, recorded in an information carrier (medium).

A piece of fixed information is only systematically accessible if it is properly documented and catalogued.

(c) Classification system for information

One important task of documentation procedures is characterising the contents, e.g. by cataloguing the information carrier into certain classes of the filing and retrieval system in use. Among the known and used classification systems are:

(1) Call number classification: characterisation by decimal number and letter combinations (e.g. Dewey Decimal System).
(2) Key-word system: characterisation by terms listed in a key-word catalogue, thesaurus, or index.
(3) Particular systems, e.g. patent classifications.

(d) Working with information

Two methods of working with information can be distinguished:

A. when studying, reading periodicals, or visiting a technical exhibition; the information is collected more or less systematically, but one should always think about problems of searching for (retrieving) the information, documenting it, and building it up into an information system.

B. when certain particular information is needed for solving a problem, one must usually:
   (a) Question ones own knowledge (memory), the results of systematic work thereby become obvious,

(b) question one's fellow workers,
(c) search in one's own library (is there a systematic filing system?),
(d) search in other data banks for various information carriers by consulting their classification systems (e.g. 53, 89),
(e) obtain a search from experts in the information service, and
(f) obtain a search from computer-resident files.

In order to perform such an information search in an effective and successful manner, every designer should have sufficient relevant knowledge about data storage and retrieval. He must:

(a) Know the available data banks and their access system,
(b) be familiar with the classification systems, and
(c) know the various types of literature, theoretical, monographs, teaching texts, etc.

Practical experience shows that it is of advantage for the designer to build up his own information system, and thereby to keep up-to-date in his own branch, and be able to find relevant information. He should do this by regular study of books and periodicals in his area. It is also possible to build up such a system for a team or group, in which case the tasks of information processing can be distributed (11).

5.5.5 Verifying/checking

For the purposes of this sub-section, we will define the terms 'verifying' in a narrow sense, as an activity usually referred to as 'checking', i.e. to determine the correctness of previous work; in a narrow sense, because one can also regard an evaluation statement such as 'the machine system conforms to the requirements' as a verification. A basic assumption must be that errors will always occur, and finds its expression in 'laws of nature' commonly attributed to (the fictitious characters) Murphy, Finagle and others. This opinion is also represented by the Method of Systematic Doubt (see Table 3), or scientific scepticism.

A *checking process* should without fail be undertaken simultaneously with the design process. Both are connected by mutually influencing procedures. A review of the design process after a longer period can either cause large losses, or result in such delays that the deadlines cannot be met. Checking is wrongly regarded as unproductive work, even though it would appear that the solution is not brought nearer by this activity. This is not true, because:

- savings occur by avoiding unnecessary work,
- valuable design capacity can be conserved,
- the follow-up activities can be performed with greater interest, based on the conviction that the previous work was not in vain.

Checking takes on various forms, and the following comments should show the relevance of the above general points to particular forms:

(1) One should assess the probability of occurrence of an error, depending on the degree of originality and complication of the problem, and on

the state of the factors, particularly of the design engineers, that influence the design process.
(2) One should assess the consequences of errors: errors in conceptualisation that are only discovered in the prototype may mean that one has to start again at the beginning, but detail errors (e.g. dimensioning a pin) can be corrected easily.
(3) Attitudes towards errors also depend on how quickly they are likely to be discovered.

General principles involved in the checking process are:
(a) None of the given data are correct,
(b) the checking path should be different from the solution path,
(c) checking should be performed after a time lapse (when self-verifying),
(d) checking should be performed by persons not involved in the solution process,
(e) assess the importance of each part for the whole, and distribute attention according to importance,
(f) assess consequences of errors, their occurrence probabilities and the possibilities of discovering them,
(g) use systematic procedures for checking,
(h) use suitable aids.

Checking takes on various forms, and the following comments should show the relevance of the above general points to particular forms:
(1) *Self-checking*: one should always check ones own work. It is recommended that this be done using methods and techniques different from those used in solving the problem, and preferably after a period of time has elapsed.
(2) Checking by a design *project* or *group leader*: the work of all designers should be regularly reviewed by the project leader with each design engineer. Technical discussions at the drawing board or desk should not serve as reviews, but are generally suitable as a vehicle for continuing education of the design engineer. An atmosphere of trust and objectivity in these discussions is very important.
(3) Review by *technical experts* (specialists): we have mentioned the breadth of design problems on various occasions. In the process it has been shown advantageous to pass certain problems on to specialists. These problems could be questions of aesthetic form-design, of ergonomics, of transport and packaging, but also of welding, casting, forging or machining problems, and in some industries (e.g. aerospace) of stress and dynamics calculations. It is important that conversations that are intended to review ideas are performed at the right time, when suggestions for changes can be incorporated without great losses.
(4) Questioning an expert also takes place in *consultancy*, but usually with the difference that a solution that is already worked out is to be evaluated, and a written report is requested.
(5) Examination by a *technical group* (syndicate) is among the most effective forms of checking. One can invite specialists from various areas; in this way many aspects can be discussed in the conference. This

procedure has been found advantageous for reviews of results in major stages (concept phase, layout), and can be used to make decisions about the subsequent steps (e.g. release for the subsequent phase).
(6) *Experiments and tests* are among the most effective review instruments, if the technical system has already been realised, e.g. as a prototype.
(7) *Modelling* is a further method for the designer to convey various types of relationships.
(8) *Value analysis* is a complex method of checking, particularly of the economic properties.

5.5.6 Representing

During the design process various physical, mathematical or conceptual relationships are needed by the design engineer to represent the machine system in certain ways: as support for visualisation, as notes for memory, or as communication with other partners. This need could concern the representation of reality at a number of levels of abstraction; they can generally be of the following types:

- *Iconic* representations that record the visualisation or the original in true form, as sketches, drawings, photographs, physical models.
- *Symbolic* representations using assumed or conventional symbols, i.e. language, mathematics, analogues.
- *Diagrammatic* representations, such as graphs, schematics, diagrams for representing relationships.

The appropriate partial tasks of representation (Fig. 10) therefore include: sketching, drawing, and modelling. The importance of representation within the design process, particularly for methodical design, has already been emphasised (DesP 11). Further details of the techniques of representation are left to the literature (1, 13, 27, 28, 47, 71, 85, 90).

The close relationship between steps of methodical design and the design documentation is evident from their information content, which includes the records of the established design characteristics and properties. Fig. 12 shows the visual analogies between the documents and the design stages shown in Fig. 11.

Chapter 6

Case Study: A Welding Positioner

Recommendation:
Please try to solve the assigned problems for each section of this chapter, prior to studying our elaboration on the following page. This procedure will tend to increase the output effect of this chapter (which acts as a technical system) on your learning process.

The design process as described and substantiated in the previous chapters should now be brought into context by means of a simple case. The selected welding positioner problem does not allow us to illustrate all shades of the spectrum of design procedures; the sequence of design documents as shown in the following pages can serve as supporting framework for understanding the recommended design procedures.
Starting situation: Attempts to increase the economic benefit are being made in all areas of technology, especially in technical processes, by using technical means. In the welding process, such improvements of efficiency and productivity can be attained from the viewpoint of handling the workpiece by using a suitable fixture (36). This should permit us to shorten the time required to obtain the desired positions of individual parts to be welded together, and to align the weld seam for best process operation; such a fixture should affect both the preparation (loading, clamping) time and the main process (active welding) time. An expected reduction of the proportion of scrap and an improvement of general weld quality are additional important benefits; the work of welding can also be eased, particularly if the weld can be performed in a down-hand position at a comfortable working height. For example, a butt-weld in 8 mm thick plate requires welding times in the ratio 10 : 11 : 17 for horizontal (down-hand) : vertical : over-head positions. Under these conditions it is understandable that the following *problem* was formulated:

To obtain the optimum weld position for a workpiece of general shape, with good accessibility from all sides.

6.1 ELABORATION OF ASSIGNED SPECIFICATION

Let us assume that the problem as assigned in the previous section has taken on the form shown in Table 5. We now need to complete this problem

TABLE 5. Welding Positioner: Assigned Problem Statement

A welding fixture is to be designed that is capable of bringing a workpiece into a desired position for welding, and to hold it in that position during work. A universally usable fixture is required, for application to a wide range of welding tasks.
Given data:
- Sizes of workpiece to be welded: max. mass 300 kg, max. base dimensions 500 × 500 mm
- Propulsion of all necessary movements to be by manual operations

---

statement, quantify as much of the data as possible, categorise the requirements, and set priorities.

1st Assigned Problem:
Elaborate the assigned specification into a clarified list of requirements (see Fig. 13, Step 1).
General hints: page 49.   Solution: Table 6.

TABLE 6. Welding Positioner: Design Specification (List of Requirements)

Questionnaire
A. Questions for necessary clarification
1. State of the art:
(a) What is the basis and sources for finding the state of the art in the area of welding positioners?
(b) Are the existing scientific-technical data and information sufficient for the design work needed on this contract?
(c) Are any research or experimentation contracts needed? Formulate the problem to be assigned.
2. Feasibility (Realisability):
Examination of the demanded welding positioner from the following viewpoints:
- technical (physical)
- economic (cost recovery, profit)
- financial (available capital, cash flow)

B. Questions to clarify the problem statement
(a) Are the given data in the contract sufficient?
(b) What other data or information are:
- to be requested from the contract sponsor (customer)?
- to be provided and prepared by the design engineer himself?
(c) Into which classes can the requirements be categorised? Differentiation between fixed requirements, minimum requirements and desires.
(d) Evaluation of the quality of the contract.

C. Methodological questions
(a) Where does this project start with respect to the general procedural plan?
(b) What preliminary decisions have already been taken before the contract was placed?
(c) Which methods and working procedures were used:
- for determining the state of the art?
- for elaborating the assigned specification, developing the list of requirements?
- for categorising the requirements?

Class of Properties (see also Fig. 5)
Function, Effect
  What should the MS do?
  What capabilities should the MS have?
Functionally Determined Properties
  Which properties are characteristic for the functions to be performed by the MS?
Operational Properties
  How well is the MS suited to the working process (operation)?
Ergonomic Properties
  How well can the human operator perform with the MS, and what influence does the MS have on the human?
Appearance Properties (Aesthetic Properties)
  What effect does the MS have on aesthetic feelings?
Distribution Properties
  How well is the MS suited to transportation, packing, and storage?
Delivery and Organisational Planning Properties
  When can the MS be delivered?
  Number to be manufactured?
Law and Standards Conformance
  Does the MS conform to standards and regulations?
Manufacturing Properties
  How suitable is the MS for manufacture?
Economic Properties
  How economic is the working and manufacturing process?
Design Properties

Table 6 (*continued*)

	Fixed Requirement	Desire
Welding Process		
• Position of the weld seam: if possible down-hand	x	
• Accessibility: weld locations over whole surface	x	
• Working height: 750 to 1200 mm above floor	x	
• Safe holding in any position	x	
Weldment		
• Material: steel, steel castings		
• Size: max. base 500 × 500 mm, max. height 1000 mm		
• Mass: max. 300 kg		
• Form: very variable, location of centre of gravity between one-third and two-thirds of workpiece height, lateral excentricity up to 100 mm		
• Connection (clamping) surface very variable, with or without holes		
• Welds at any location on the weldment surface		
• Temperature: hot, locally very high temperature		
Working Conditions		
• Dirty operation and environment		
Application and Maintenance		
• Frequency of use: very high		
• Demanded life: minimum 5 years	x	
• Maintenance requirement:		
minimum	x	
none		x
• Requirement for Floor: concrete (reinforced)		
Requirements regarding operation and safety		
• Stiffness and stability		
• Safe holding (danger of tilting)	x	
• Easy and safe adjustment and movement (accessibility of drive locations)	x	
• No injury danger	x	
Appearance of the apparatus		
• Pleasing form		x
• Surface quality		
insensitive to damage		x
easy to clean	x	
neutral colour		
Manufacture		
• Small batch		
Price		
• Competing product about £250		
Transport and storage		
• Minimum storage space		x
• Change of location		
with crane	x	
without crane		x
Patent situation		
• No patent violation	x	
Standards and regulations		
• No particular regulations		

Case Study: A Welding Positioner   81

Following the discussions presented in Sections 5.4 and 5.5, p. 48 ff. the list of requirements can be similar to that shown in Table 6. A *systematic analysis* of individual viewpoints can be observed. Even if the sequence has not been strictly maintained, the background of the *classes of properties* numbers 1 to 10 in Fig. 5 is noticeable. If these groups are studied carefully, it will be clear that many of the properties can be placed into the class of functional parameters (class 2) and into that of operational properties (class 3). This seems reasonable for a manufacturing system. The priority rating of the properties is divided into two levels, fixed requirements and desires.

6.2 ESTABLISHING THE FUNCTIONAL STRUCTURE

2nd Assigned Problem:
Now establish the technical processes (steps 2.1 to 2.3).
General hints: pages 50, 51, and Fig. 2.
Particular hints: it is useful in this case to show the execution and conclusion phases within the TP. The boundaries of the 'black box' process, and of the technical process, are chosen to follow different lines.
Solution: Fig. 18.
General form of solution page, see below.

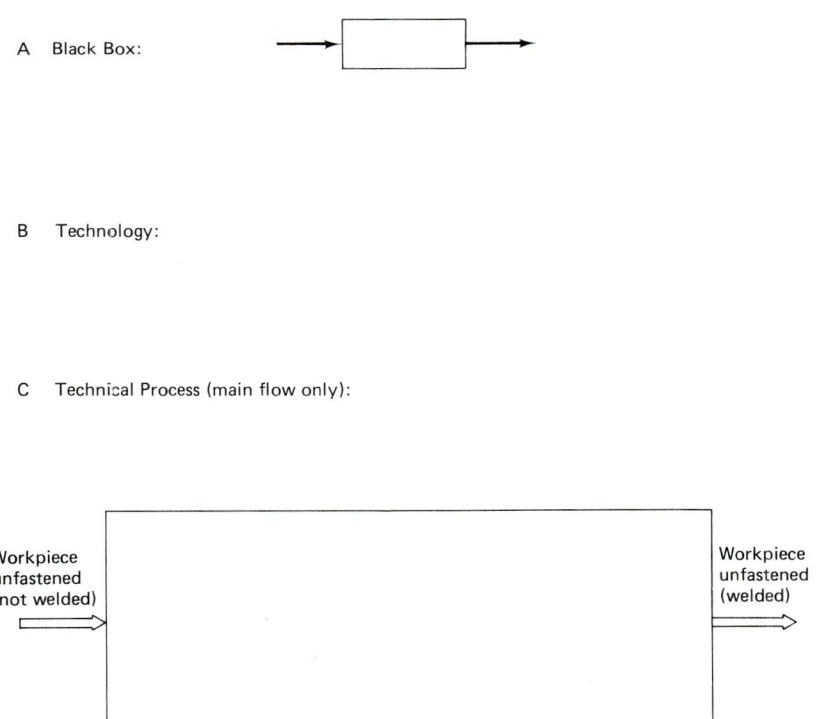

A   Black Box:

B   Technology:

C   Technical Process (main flow only):

Workpiece
unfastened
(not welded)

Workpiece
unfastened
(welded)

Fig. 17

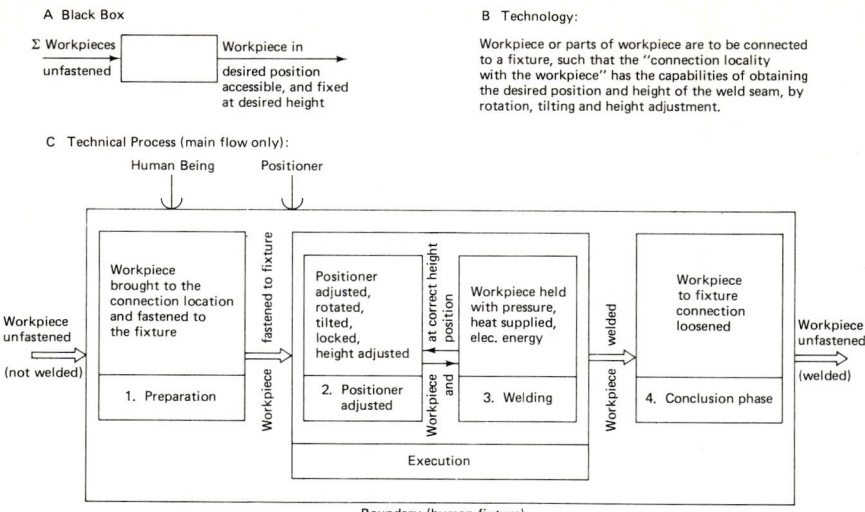

Fig. 18 Welding Positioner: Technical Process (Block Schematic).

By abstraction, we obtain the core of the problem, as shown in Fig. 18, section A. Even if the question about the *technology* appears surprising in this context, it is justified. From the few available possibilities, the following facts have been established for this case: workpieces are fastened to a fixture (the positioner); the connection locality between fixture and workpiece has all movement capabilities to attain the desired position, direction and height for the weld to be performed. We could remain at this level, and defer further decisions about the nature of these capabilities until considering the functional structure, and the search for function-carriers. According to Fig. 18, section B, decisions have been made on the basis of a motion analysis (degrees of freedom) and of experience, to establish these capabilities as: rotation about a vertical axis, tilting around a horizontal axis through 90° (or better ± 90°), and a height adjustment.

In this case, consideration of the *technical process* can almost be neglected, because it is already represented by an extensive description of the technology, in which we have explicitly mentioned the partial functions that we seek. If alternatively the technical process had been shown only as its main flow, following the pattern of Fig. 2, an almost three-dimensional picture of the process would have emerged. The list of requirements can thereby be expanded by a few more statements. For this purpose, an extension of the process in the region of the conclusion phase would be very advantageous.

3rd Assigned Problem:
Now determine the functional structure (Steps 2.4 to 2.7).
General hints: pages 51–53, Fig. 18, section C.
Particular hints: the functional structure can be represented in two forms: block schematic, and hierarchical function tree.
Solution: Figs 20 and 21.
General form of solution page, see page 83.

Case Study: A Welding Positioner 83

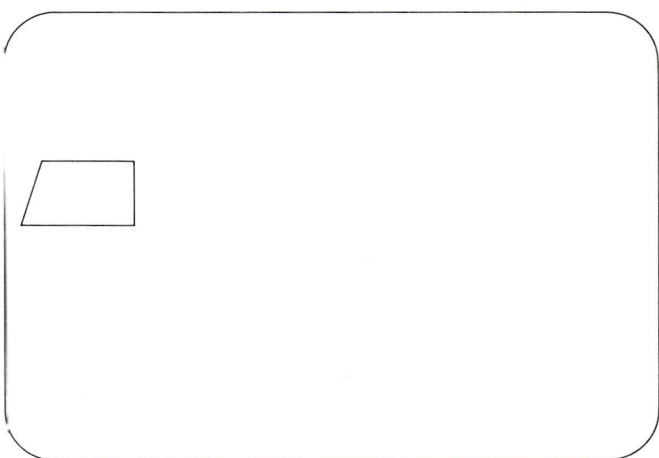

Fig. 19

Case Study: A Welding Positioner

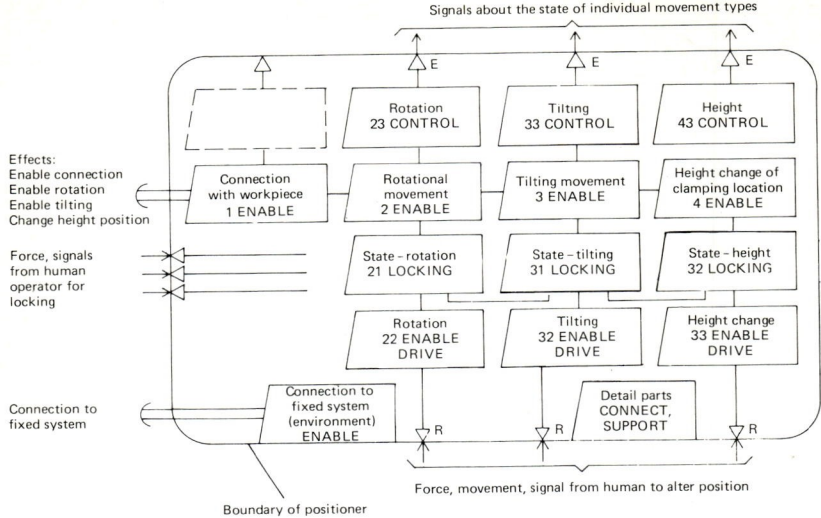

Fig. 20 Welding Positioner: Functional Structure (Block Schematic).
R: Receptor, E: Effector

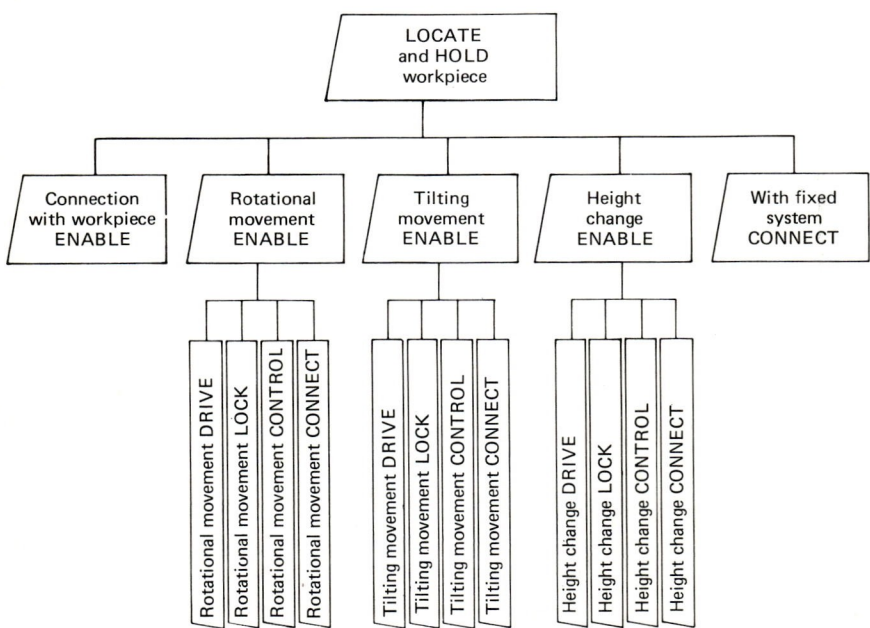

Fig. 21 Welding Positioner: Functional Structure (Hierarchical Tree).

Case Study: A Welding Positioner 85

The concept principle should be assembled from as few of the organs as possible. The other organs can be added to the selected variant at a later stage. The ruling condition is that these added organs must be compatible with all variants.
Solution: Figs 23 and 24.
General form of solution page, see below.

Partial Functions	Action Principles/Classes of Function-carriers					
	1	2	3	4	5	...

Fig. 22

The next step is a derivation and completion according to Fig. 1. One can now clearly detect the boundaries of the positioner, and of its receptors, effectors, and input values in the result shown in Fig. 20.

6.3 ESTABLISHING THE CONCEPT

In our simple fixture, all inputs arise from the human being: force, movement, and signals. We must later check that the ergonomic parameters of the human operator have not been exceeded, e.g. whether the table is too heavy. The search for action principles, and families of function-carriers, can now take place in a *morphological matrix*.

4th Assigned Problem:
For each of the functions shown in the functional structure, find possible action principles, and classes of function-carriers (Steps 3.1 and 3.2). Then combine these individual organs into a total system (Steps 3.3 and 3.4). Following this, evalute a few variants (Step 3.5).
General hints: pages 54–56; evaluation: pages 68–71 (point rating).
Particular hints: some of the functions occur more than once in the functional structure. They are entered only once into the morphological matrix.

Case Study: A Welding Positioner

Partial Functions		Action Principles/Families of Function-carriers					
		1	2	3	4	5	6
1	ENABLE connection with workpiece	Form interlocking	← mechanical → screw or bolt / wedge		← Force locking (friction) → pneumatic / hydraulic / magnetic		
2	ENABLE rotational movement	Rotational guidance, sliding journal bearing	rolling bearing				
3	ENABLE tilting movement	cylinder	sphere	fulcrum pin position	hang from above		
4	ENABLE height adjustment	straight line guidance, bearing, sliding or rolling	screw thread		lever mechanism		
21 31 41	LOCK state	hold directly by hand, weight of workpiece	← form interlocking → hole – pin / ratchet mechanism		← force locking (friction) → within guidance / screw screw with washer / wedge, brake block		
22 32 42	DRIVE (by hand)	Direct	← with mechanical advantage device → gear wheel pair / rack and pinion / helical gears (crossed) / worm and worm-wheel / band, rope, chain / lever, eccentric cam				
23 33 43	CONTROL of movement	through drive mechanism	through locking (ratchet)				
	Show position	← mechanical → line scale / pointer scale		optical	electronic	mechanical stop	

Fig. 23 Welding Positioner: Morphological Matrix.

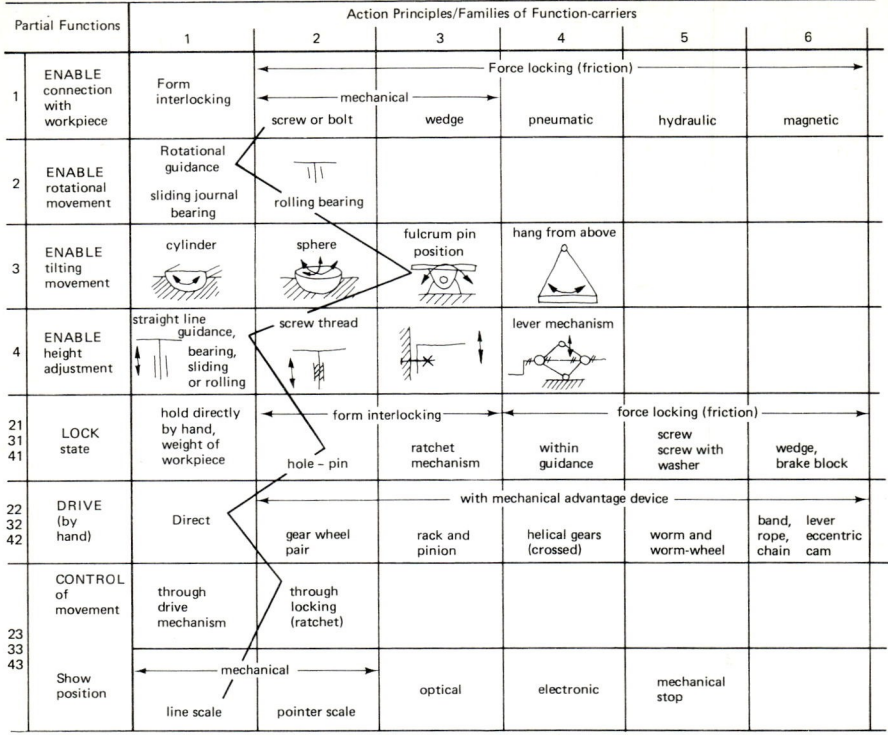

Fig. 24 Welding Positioner: Concepts and Concept Sketches.

Case Study: A Welding Positioner

Our example deliberately contains very differing results for the individual functions shown in Fig. 23, especially concerning the degrees of concretisation of proposals, just as would be expected in a real solution process.

Further considerations permit the conclusion that the functions of connection, 1, and also of locking, propulsion, and control, can be combined in any desired way with the work functions 2, 3, and 4. The optimal overall solution is thus obtained by an optimal combination of these functions. Fig. 24 shows only four sample solutions, in which we have selected to use the same family for the functions 2 and 4 (namely solution 1 in Fig. 24), and only the means of tilting are varied.

Technical evaluation shows an almost equal rating for these solutions, even though some show greater promise with respect to locking and control. An estimation of costs of manufacture by comparing only the commonest parts permits us to select alternative D as the most favourable.

6.4 ESTABLISHING THE PRELIMINARY LAYOUT

Starting from alternative D, we recognise that there is a potential weakness in the possibilities of locking the tilting motion. A relatively large tilting moment would have to be held on a relatively small diameter. Therefore one should now concentrate on finding a better solution for this function, as well as complying with the usual considerations of the design process schematic, Fig. 13; in this case, an additional guidance on a larger radius is desirable. Calculations with respect to statics and mechanics of materials should reveal minimum dimensions, especially considering the limitations imposed by the capabilities of the human operator. The other function-carriers can be selected by optimising within each row of the morphological matrix; their compatibility with other solutions was a prior condition.

5th Assigned Problem:
Starting from conceptual alternative D, the form of the positioner should now be developed in a preliminary layout sketch (Steps 4.1 to 4.5).
General hints: pages 57 – 59.
Solution: Fig. 25.
The result of this step, the preliminary layout shown in Fig. 25, applies a number of different function-carriers selected from the morphological matrix, Fig. 23, especially for locking the various motions.

6.5 ESTABLISHING THE DIMENSIONAL LAYOUT

Further concretisation of all design properties, starting from the preliminary layout, can lead to a dimensional layout as shown in Fig. 26. With additional optimisation and review, this layout can serve as the starting point for preparing the detail drawings, a procedure that is left to the reader.

Case Study: A Welding Positioner

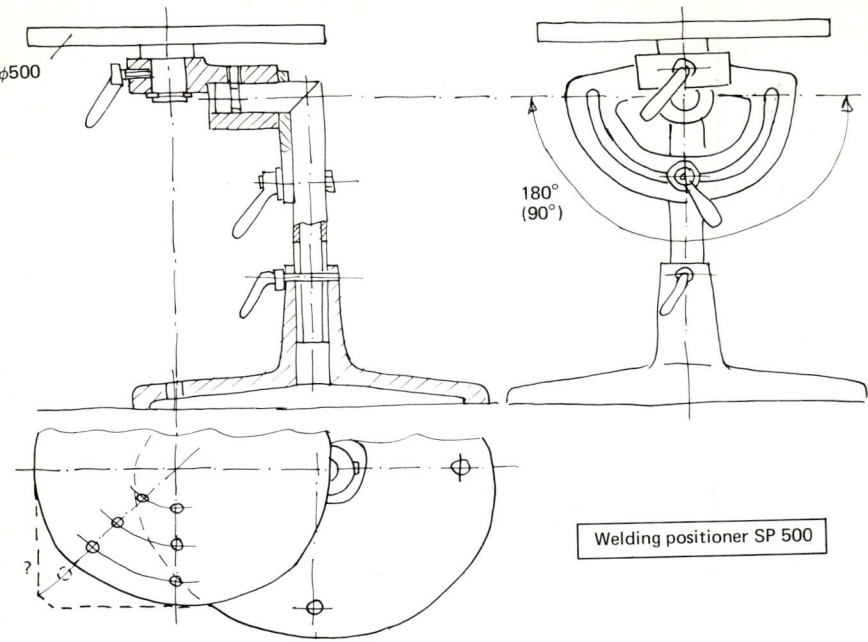

Fig. 25 Welding Positioner: Preliminary Layout as Hand Sketch.

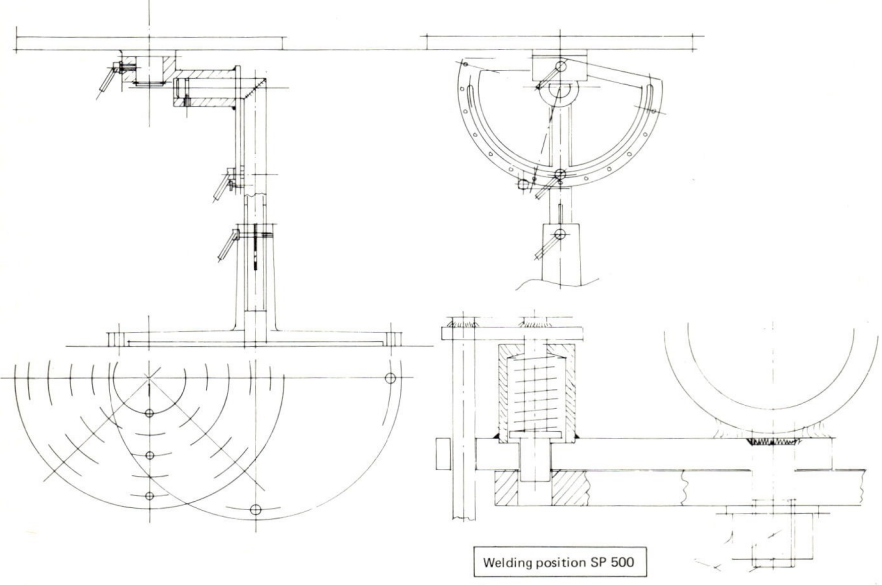

Fig. 26 Welding Positioner: Dimensional Layout with Detail.

Chapter 7

Summary

A design theory that can be regarded as a strategy for problem-solving in design engineering must respect two areas of knowledge: the theory concerning machine systems as objects of design work, and the theory of the design process itself. For this reason, this book first seeks to explain technical processes, to view them as systems of transformations to attain desired states of throughput materials, energy and information; then it seeks to treat machine systems as deliverers of the necessary input actions to the process, in order that it can perform those transformations.

The subsequent discussions deal with the structure, mode of action, properties, origination, development, and systematics of such technical systems. Further, the design process is analysed with respect to assigned problems, modelling, structure, strategies, tactics, representation, and working means, in order to describe in detail the general model of methodical procedure which forms the main subject of this book.

The separate design steps are treated in a unified fashion from various viewpoints (entrance, exit conditions, objectives, established characteristics of the technical system, and suitable methods to aid the designer), and is accompanied by a case study showing the significant stages of design of a welding positioner.

Chapter 8

Bibliography

This bibliography is placed in alphabetical order within two sections. The items numbered 1 to 36 constitute the literature section of the original German text. Of these, items 18, 20, 26, 28 and 30 are English-language works, or are available in translations as noted. The remaining references have been added for this English edition, either to provide similar information, or to offer works that expand on some sections of this book; these extensions reach into topics of management, detail design, manufacture, etc.

(1) Andreasen, M. M., 'Darstellungsmöglichkeiten beim Konzipieren,' *Schweizer Maschinenmarkt,* Bd. 78, H. 4 and 6, 1978

(2) Beitz, W., 'Übersicht über Konstruktionsmethoden,' *Konstruktion,* Jg. 24, H. 2 and 3, 1972

(3) Ewald, O., *Lösungssammlungen für das methodische Konstruieren,* VDI-Verlag, Düsseldorf, 1975

(4) Hansen, F., *Konstruktionssystematik,* 2nd edn, VEB-Verlag Technik, Berlin, 1966

(5) Hansen, F., *Konstruktionswissenschaft,* Carl Hanser Verlag, München, 1974

(6) Heinrich, W., *Optimale Konstruktionsverfahren ermöglichen eine Rationalisierung des Konstruktionsprozesses,* Manuscript, 1967

(7) Heinrich, W., 'Eine systematische Betrachtung der konstruktiven Entwicklung technischer Erzeugnisse,' *Maschinenbautechnik,* Nr. 5, 1973, to Nr. 12, 1976

(8) Hubka, V., *Theorie der Maschinensysteme,* Springer-Verlag, Berlin, 1973

(9) Hubka, V., *Theorie der Konstruktionsprozesse,* Springer-Verlag, Berlin, 1975

(10) Hubka, V., 'Konstruktionswissenschaft,' *VDI-Zeitschrift,* Jg. 116, H. 20, 22 and 24, 1974

(11) Hubka, V., 'Der Konstrukteur als Informationsverbraucher,' *Schweizer Maschinenmarkt,* Bd. 73, H. 38, 40 and 42, 1973

(12) Hubka, V., 'Intuition und Konstruktionsgefühl,' *Schweizer Maschinenmarkt,* Bd, 75, H. 50, 1975

(13) Hubka, V., 'Darstellen und Modellieren beim Konstruieren,' *Schweizer Maschinenmarkt,* Bd. 76, H. 33, 35 and 37, 1976

(14) Hubka, V., 'Der grundlegende Algorithmus für die Lösung von Konstruktionsaufgaben,' *TH Ilmenau,* 1967
(15) Hubka, V., 'Bewerten und Entscheiden beim Konstruieren,' *Schweizer Maschinenmarkt,* Bd. 74, H. 20, 22 and 24, 1974
(16) Kesselring, R., *Bewertung von Konstruktionen,* VDI-Verlag, Düsseldorf, 1951
(17) Koller, R., *Konstruktionsmethode für den Maschinen-, Geräte- und Apparatebau,* Springer-Verlag, Berlin, 1976
(18) Leyer, A., *Maschinenkonstruktionslehre,* Birkhäuser-Verlag, Basel, H.1, 1963; H.2, 1964; H.3, 1966; H.4, 1968. English edition: *Machine Design,* Blackie, London, 1974
(19) Lowka, D., 'Über Entscheidungen im Konstruktionsprozess,' Thesis D 17, *TH Darmstadt,* 1976
(20) Morrision, D., *Engineering Design,* McGraw-Hill, New York, 1969
(21) Ott, H. H., 'Konstruktionsmethodik. Stichworte zur Vorlesung,' *ETH Zürich,* 1975
(22) Pahl, G. and Beitz, W., *Konstruktionslehre,* Springer-Verlag, Berlin, 1977
(23) Rodenacker, W. G., *Methodisches Konstruieren,* 2nd edn, Springer-Verlag, Berlin, 1970
(24) Roth, K., 'Aufbau und Verwendung von Katalogen für das methodische Konstruieren,' *Konstruktion,* Jg. 24, H. 11, 1972
(25) Roth, K., Franke, H. J. and Simonek, R., 'Algorithmisches Auswahlverfahren zu Konstruktion mit Katalogen,' *Feinwerktechnik,* Jg. 75, Nr. 8, 1975
(26) Shigley, J. G., *Mechanical Engineering Design,* McGraw-Hill, New York, 1963
(27) Tjalve, E. and Andreasen, M. M., 'Zeichnen als Konstruktionswerkzeug,' *Konstruktion,* Jg. 27, H. 2, 1975
(28) Tjalve, E., *Systematische Formgebung für Industrieprodukte,* Fachpresse Goldach, Goldach, 1978
Original edition in Danish: *Systematisk udformning af Industriprodukter,* 1976
English edition: *Short Course in Industrial Design,* Newnes-Butterworths, London, 1979
(29) Wörgerbauer, H., *Die Technik des Konstruierens,* Oldenbourg-Verlag, München, 1943
(30) Woodson, T. T., *Introduction to Engineering Design,* McGraw-Hill, New York, 1966
(31) VDI-Richtlinie 2210 (Entwurf), Blatt 1: 'Datenverarbeitung in der Konstruktion, Analyse des Konstruktionsprozesses im Hinblick auf den EDV-Einsatz,' VDI-Verlag, Düsseldorf
(32) VDI-Richtlinie 2222: 'Konstruktionsmethodik, Konzipieren technischer Probleme, Produkte,' VDI-Verlag, Düsseldorf, 1976
(33) VDI-Richtlinie 2225, Blatt 1 and 2: 'Technisch-wissenschaftliches Konstruieren,' VDI-Verlag, Düsseldorf, 1975
(34) VDI-Richtlinie 2802: 'Wertanalyse, Vergleichsrechnungen,' VDI-Verlag, Düsseldorf, 1971
(35) VDI-Bericht 219: 'Konstruktion als Wissenschaft,' VDI-Verlag, Düsseldorf, 1975

(36) *Vorrichtungen und Arbeitschilfen für Schweissarbeiten,* Merkblatt 149, 4th edn, Beratungsstelle für Stahlverwendung, Düsseldorf, 1970
(37) Adams, J. L., *Conceptual Blockbusting,* Freeman, San Francisco, 1974
(38) Allan, J. J., *C.A.D. Systems,* North-Holland, Amsterdam, 1977
(39) Archer, L. B., *Technological Innovation—A Methodology,* Inforlink, Frimley, Surrey, 1971
(40) Ashford, F., *The Aesthetics of Engineering Design,* Business Books, London, 1969
(41) Asimow, M., *Introduction to Design,* Prentice-Hall, Englewood Cliffs, N.J., 1962
(42) Bailey, R. L., *Disciplined Creativity for Engineering,* Ann Arbor Science, Ann Arbor, Mich., 1978
(43) Besant, C. B., *Computer Aided Design and Manufacture,* Horwood, Chichester, Sussex, 1980
(44) Bjørke, Ø., *Computer-Aided Tolerancing,* Tapir, Trondheim, 1978
(45) Booker, P. J. (ed.), *Conference on the Teaching of Engineering Design,* Inst. Engng. Des., London, 1964
(46) Booker, P. J. (ed.), *Teaching Engineering Design,* Inst. Engng. Des., London, 1966
(47) Buck, C. H., *Problems of Product Design and Development,* Pergamon, Oxford, 1963
(48) Cain, W. D., *Engineering Product Design,* Business Books, London, 1969
(49) de Bono, E., *The Use of Lateral Thinking,* Penguin, Harmondsworth, 1971
(50) de Bono, E., *Teaching Thinking,* Temple Smith, London, 1976
(51) Eder, W. E. and Gosling, W., *Mechanical System Design,* Pergamon, Oxford, 1965
(52) Eder, W. E., *Methodology, Techniques and Counselling in Design Teaching,* Prace Naukowe Instytutu Cybernetyki Techniczej, Politechniki Wrocławskiej, No. 54, vol. Konferencje, No. 19, 1978, pp. 197–210
(53) *E.S.D.U. Index 1979–80,* Engineering Sciences Data Unit, London, 1979
(54) Flursheim, C., *Engineering Design Interfaces,* Design Council, London, 1977
(55) French, M. J., *Engineering Design: The Conceptual Stage,* Heinemann, London, 1971
(56) *The Fulmer Materials Optimizer (2nd ed.),* Fulmer Research Inst., Slough, Bucks., 1979 (also General Electric Co., Schenectady, N.Y.)
(57) Gage, W. L., *Value Analysis,* McGraw-Hill, London, 1967
(58) Glegg, G. L., *The Design of Design,* Cambridge Univ. Press, Cambridge, 1969
(59) Gordon, W. J. J., *Synectics,* Harper & Row, New York, 1961
(60) Gregory, S. A. (ed.), *The Design Method,* Butterworths, London, 1966
(61) Gregory, S. A. (ed.), *Creativity in Engineering,* Butterworths, London, 1970

(62) Harrisberger, L., *Engineersmanship, A Philosophy of Design,* Brooks/Cole, Belmont, Calif., 1966
(63) Heller, E. D., *Value Management: Value Engineering and Cost Reduction,* Addison-Wesley, Reading, Mass., 1971
(64) Johnson, R. C., *Optimum Design of Mechanical Elements (2nd ed.),* Wiley, New York, 1980
(65) Jones, J. C., *Design Methods: seeds of human futures,* Wiley, New York, 1970
(66) Kirk, N. S. and Ridgway, S., 'Ergonomics Testing of Consumer Products,' *Applied Ergonomics,* **1**, Dec. 1970, pp. 295 – 300, and **2**, Mar. 1971, pp. 12 – 18
(67) Krick, E. V., *An Introduction to Engineering and Engineering Design,* (1st edn), Wiley, New York, 1965. (Also 2nd edn, 1969, but omitting referred figures)
(68) Layard, R. (ed.), *Cost-Benefit Analysis,* Penguin, Harmondsworth, 1972
(69) Leech, D. J., *Management of Engineering Design,* Wiley, London, 1972
(70) Lent, D., *Analysis and Design of Mechanisms* (2nd edn), Prentice-Hall, Englewood Cliffs, N.J., 1970
(71) Liggett, J. V., *Fundamentals of Position Tolerance,* Soc. Man. Eng., Dearborn, Mich., 1970
(72) Marples, D. L., 'The Decisions of Engineering Design,' *Inst. Engng. Des.,* London, 1960
(73) Matchett, E., 'The Controlled Evolution of Engineering Design,' *Inst. Engng. Des.,* London, 1963
(74) Matchett, E., 'Work Study in Machine and Product Design,' *Work Study and Management,* Jan. 1964, pp. 21 – 28
(75) Matousek, R., *Engineering Design: A Systematic Approach,* Blackie, London, 1963; German original edition: *Konstruktionslehre des allgemeinen Maschinenbaues,* Springer, Berlin, 1957
(76) Mayall, W. H., *Industrial Design for Engineers,* Iliffe, London, 1967
(77) Meller, J., *The Buckminster Fuller Reader,* Penguin, Harmondsworth, 1972 (Pelican imprint)
(78) Middendorf, W. H., *Engineering Design,* Allyn & Bacon, Boston, Mass., 1969
(79) Mudge, A. E., *Value Engineering,* McGraw-Hill, New York, 1971
(80) Osborn, F. E., *Applied Imagination,* Scribner's, New York, 1953
(81) Papanek, V., *Design for the Real World,* Bantam, Toronto, 1973
(82) Peck. H., *Allocating Tolerances and Limits,* Longmans, London, 1963
(83) Pitts, G., *Techniques in Engineering Design,* Butterworths, London, 1973
(84) Polya, G., *How to Solve It,* Princeton Univ., Princeton, N.J., 1945
(85) Prince, D. M., *Interactive Graphics for C.A.D.,* Addison-Wesley, Reading, Mass., 1971
(86) Roark, R. J. and Young, W. C., *Formulas for Stress and Strain* (5th edn), McGraw-Hill Kogakusha, Tokyo, 1975
(87) Roe, P. H., Soulis, G. N. and Handa, V. K., *The Discipline of*

Design, Allyn & Bacon, Boston, Mass., 1967
(88) Ruiz, C. and Koenigsberger, F., *Design for Strength and Production,* Macmillan, London, 1970
(89) Siddall, J. N., *Mechanical Design Reference Sources,* Univ. of Toronto, Toronto, 1967
(90) Tjalve, E., Andreasen, M. M. and Schmidt, F. F., *Engineering Graphic Modelling: workbook for design engineers,* Newnes-Butterworths, London, 1979
(91) Thring, M. W. and Laithwaite, E. R., *How To Invent,* Macmillan, London, 1977
(92) Tuttle, S. B., *Mechanisms for Engineering Design,* Wiley, New York, 1967
(93) Vernon, P. E., *Creativity,* Penguin, Harmondsworth, 1970
(94) Wallace, P. J., *The Techniques of Design,* Pitman, London, 1952
(95) Wearne, S. H., *Principles of Engineering Organisation,* Arnold, London, 1973
(96) Wickelgren, W. A., *How to Solve Problems,* Freeman, San Francisco, 1974
(97) Woodson, W. E. and Conover, D. W., *Human Engineering Guide for Equipment Designers* (2nd edn), Univ. of California Press, 1970
(98) Zwicky, F., *The Morphological Method of Analysis and Construction,* Courant Anniv. Vol., Wiley-Intersc., New York, 1948
(99) BS 5750:1979: *Quality Systems,* British Standards Institution, London, 1979

WDK-Series, Heurista, Zurich:

WDK 1: Hubka, V., *Allgemeines Vorgehensmodell,* 1980
WDK 2: Hubka, V., Hora, S. and Andreasen, M. M., *Bibliographie des Konstruktionsgebietes* (Bibliography of Design Science), 1981
WDK 3: Hubka, V., Andreasen, M. M., Eder, W. E., Pighini, U., Schlesinger, A. and Wyss, M., *Fachbegriffe der wissenschaftlichen Konstruktionslehre in 6 Sprachen* (Terminology of the Science of Design Engineering in 6 languages), 1981
WDK 4a: Andreasen, M. M. and Hubka, V., *Methodisches Konstruieren von Maschinensystemen – Fallbeispiele,* 1981
WDK 5: Hubka, V. (ed.), *Konstruktionsmethoden in Übersicht* (Review of Design Methods), 1981
WDK 6: Hubka, V. (ed.), *Konstruktionsunterricht in Übersicht* (Review of Teaching Engineering Design), 1981
WDK 7: Hubka, V. and Eder, W. E. (eds.), *Ergebnisse der 'ICED 81' Rom* (Conference Results—'ICED 81' Rome), 1981

Chapter 9

Glossary

This glossary summarises some important ideas and terms used in this book, and gives brief examples. Wherever possible, the meanings of words and expressions has been checked against the Concise Oxford Dictionary (6th edn), in order to apply a preferably accurate and consistent usage throughout this book, even if this does not always coincide with current colloquial and technical English. I have also attempted to select the most appropriate translation of meanings, by consulting the author, by checking against Langenscheidt's New Concise German Dictionary (1973 edn), and Roget's Thesaurus (Penguin, 1966 edn).

9.1 NARRATIVE DEFINITIONS

This section provides some preliminary definitions, set in a colloquial language to outline the thinking of the main chapters. Better definitions, with more accurate terminology and more consistent usage, may be found in the synopsis sections of Chapters 2 to 4, and in the second section of this glossary. This set of statements is deliberately broad and non-specific, and contains some words used in their common meanings, but also some for which the meaning is restricted to comply with the intentions of this book.

(a) At an *abstract level* of considering a device:
- *system* – a collection of tangible and intangible entities (components) that are connected together in a defined way, and shows a predictable behaviour. It is also a way of organising such components into a meaningful whole. This interpretation is thus broader than the usual English-language outlook.
- *process* – something (intangible) that works over a period of time to transform (from input to output) a quantity of materials, energy, and information, into different, preferably better and more usable, forms. It is influenced by the system within which it works.
- *structure* – a perceived form of internal organisation which characterises the collection of components into a system, and their interactions among themselves and with the environment. This term should not be confused with the common usage of the word that

means a material framework, for example one that supports a traffic overpass.
- *form* – a perceived external organisation, including shape, geometry, general appearance, aesthetic impression, etc. A change of form may simultaneously involve a change of structure.
- *hierarchy* – a meaningful arrangement in an ascending order of abstractions (each constituting a level of abstraction), of groupings of similar systems or processes (at any one level). Higher abstractions contain the lower ones, have fewer members, but each member group contains more individuals, and there is larger variation in properties, characteristics and features between members of a group. The levels are designated in a similar way to biological levels: phylum, class (the most-used word in this sequence), family, genus, species or type, and individual. When working in the continuum between abstract and concrete, moving into a higher level of the hierarchy is termed *abstracting*, moving into a lower level is termed *concretising*.
- *component* – an entity, either tangible or abstract, that is used as a contributed part to a structure or a system, and helps to perform the actions and effects needed to bring about a process. It can exist on its own, but is then usually not capable of performing all its duties.

(b) At a *concrete level* of considering a device:
- *function* – that assigned problem which a system must solve, namely the effects that are exerted by that system onto another mechanism, the process, or the operand within the process.
- *effect* – the action of one system, mechanism or object on to another. The object that has the effect exerted on it reacts by changing its properties, or its state of existence.
- *property* – something possessed by an object, effect or function that can be measured on an agreed scale to permit a statement of a value and the unit of measurement. The abstract property without measurement or value is a *characteristic*, or a feature.
- *operand* – something tangible, consists of materials, energy and information, and is the input, throughput, and output of a process; it is transformed within that process.
- *organ* – a combination of items that perform a function, and thereby facilitate a process. Just as in biology, an organ can be a separable entity (e.g. the heart), or a set of connected parts of other entities (e.g. the knee joint). In general, an organ will permit certain motions, and will restrict others, i.e. it limits the available degrees of freedom, preferably in the best way to achieve the organ's function. It is thus a particular form of *function-carrier*.
- *constructional element* (machine element, detail part, etc.) – a separately manufacturable item, that is assembled with others into a tangible, physical mechanism, and that carries, supports, connects, transmits (e.g. forces), etc.
- *mechanism* – something tangible that by means of the function of its organs produces a physical effect on its own environment, and

thereby implements a process that is neighbouring part of its environment. A mechanism must therefore communicate with an operand according to a definite pattern (by means of effectors and receptors), such that a desired process is performed. Two distinct kinds of property co-exists: that of *functioning* or of performing, and that of *existing* or of having been manufactured. Each kind of property needs its own type of *structure* to define it and its workings. These physical and mental structures (and the ideas behind them) are caused to merge into one another by the processes of manufacture and assembly, so that the final device contains both in an indivisible way.

(c) From the viewpoint of the *human being* faced with a need:
- *planning* – the mental process of imagining ways in which the need can be fulfilled, deciding on the most advantageous way under the circumstances, and preparing for the next step. It is performed before an object or organisation can exist, it is a pre-requisite to that existence. *Design*, and especially design engineering, is a special form of planning. In this book, the author suggests that functions and organs can be established at a more abstract level of thought, i.e. before the mechanisms and constructional elements are defined.
- *requirements* – statements about the values of properties that are deemed to be acceptable for the projected artefact.
- *establishing* – the process of generating possible alternatives, and selecting between them, in order to produce the description of the proposed elements. An object or procedure needs to be *established* before it can exist; if an object or procedure exists, its properties can be *determined*; the usage of words is somewhat arbitrary, and can be *defined*.
- *concept* – a mental (and usually somewhat abstract) model of a physical reality, e.g. of a mechanism, or a set of circumstances.
- *realising* – the process of transforming the developed ideas into a reality, making it exist as a mechanism or organisation. Putting material around a mental framework, or *embodiment*, is a necessary part of this step, but so is manufacturing at a more concrete level of consideration.
- *implementation* – the final step needed to make the artefact (physical, mental or organisational) work. It usually also requires continuing supervision and management.
- *control* – the process of looking after things (and ideas), so that they do not get out of hand, and are preferably steered in the most favourable direction. It can be thought to consist of two stages: regulation as the process of aiming in a predicted direction at the start of an operation, and feedback to correct that aim from time to time. In terms of design operations, *verifying* is the regulating action of establishing a prediction of truth, *checking* is the feedback action of detecting errors.

9.2 KEY-WORD DEFINITIONS

This section collects some formal definitions of terms used in this book, placed in alphabetical order of important key-words. It is intended to complement the main text of the chapters, but is not cross-referenced to them. Within each entry, other terms referenced are set in italics, e.g. *action organ*, and the item number is appended, as (3).

1. Action locality (also action space, surface, line, point)

That part of a function-performing mechanism (the *action organ* (3)) that directly mediates an active or passive *effect* (19) (communicates or transmits an action or *motion* (2), or performs a function). It is always located at the *boundary* (45) of (the relevant portion of) the *machine system* (30) under investigation (effector as output action locality, receptor as input locality to the system). The action locality is defined by the *design properties* (16) (dimensions, material and surface properties, etc., of the component parts). Examples: (1) a machine tool lead screw centre (action line), (2) outer surface of a car brake shoe lining (action surface), (3) combustion chamber volume of an internal combustion engine (action space).

2. Action motion

Motion that is necessary for the *effect* (19) to be achieved, or the *function* (25) to be performed, and that is realised (brought about) by the *organs* (34) of the *machine system* (30). The action motion may be rotary or linear (or any combination of these in three-dimensional space), continuous or reciprocating, deterministic or random, etc. An action motion is not necessarily a mechanical/geometric/kinematic motion, other *action principles* (5) demand different transformations (e.g. chemical) that are nevertheless termed 'motions'. Examples: (1) working stroke of a shaper (linear reciprocating), (2) rotation of a lathe spindle (rotary continuous), (3) movement capability of a car suspension shock absorber (internally linear random, externally rotational).

3. Action organ

That portion of a *machine system* (30) that brings about the action, as effector or receptor, at the *boundary* (45) of the machine system. An action organ performs a self-contained *function* (25), and does not necessarily coincide with a physical sub-assembly of parts. The *action localities* (1), spaces, surfaces, lines, or points are the essential attributes of the action organ. Example: a clutch lining, plus pressure plate, plus springs, plus force reaction housing.

4. Action pair

An *action organ* (3) consisting of two interacting items between which the *effect* (19) takes place; it usually consists of an active and a passive element of the pair. Example: (1) clutch plate hub internal splines, and shaft splines, (2) meshing gear tooth profiles.

5. Action principle

Natural phenomenon of physical, chemical or biological nature on the basis of which an *effect* (19) can be achieved. An action principle can combine a number of *modes of action* (32). Example: oxidation of a material (chemical action principle), and expansion of a gas (thermodynamic principle), can be combined (1) within an internal combustion engine, (2) for an explosive charge.

6. Classifying aspect

Statement of essential *purpose* (39) (key task) that needs to be achieved, important qualitative design *factor* (21) from which a number of possible solutions can be systematically developed by changing *design properties* (16) or *action principles* (5).

7. Degree of abstraction

The position of the state of development of a *machine system* (30) (MS) in its progress from 'abstract' to 'concrete'. Degree of abstraction for *functions* (25) can be defined in a similar way with respect to required or performed functions. The opposite to abstraction is termed *concretisation*. Example: (1) a 'stepwise variable speed power transmission' is an abstract designation for a machine system family, a 'change-speed gear box' is a more concrete designation for a MS genus, a '4-speed 3-shaft synchromesh gear box of 50 kW capacity' is a concrete designation for a single type of machine, see Table 2 and Fig. 8; (2) 'screw' is an abstract designation for all types of screws, a concrete designation for a single type of screw is M 16 × 50, ISO hexagon socket head, high tensile steel.

8. Degree of complication of the machine system

A measure of the complexity of a *machine system* (30) (usually as number of components), or of *functions* (25) or other characteristics. Basically divided into four hierarchical grades; e.g. with respect to number of parts: Components, Assemblies, Machines, and Equipments (or Plants). At each level a finer division is possible. *Note*: the word 'complication' is intended to describe a reliability-related characteristic resulting from number of parts. 'Difficulty' (e.g. of manufacture) and 'intricacy' (e.g. of component form) are used in relation to the *form* (22) of components and assemblies, with the intention of avoiding the word 'complexity' that includes both meanings.

9. Degree of intricacy (also of difficulty)

Characteristic of the technological intricacy, difficulty of manufacture, involvement, complexity of realisation of MS sub-groupings, within particular *degrees of complication* (8) of the *machine system* (30). It is possible to classify machine components by degrees of intricacy: from a simple pin or washer, to the most complicated casting or forging. Different systems use four or more classes (typically seven). The more complicated systems of classification can describe a 'family' of sufficiently similar parts suitable for manufacture within a 'group technology' arrangement of machines.

10. Degree of originality of the machine system

Originality is in this context defined as the relative content of novel *principles* (35) or components in the *machine system* (30) MS. One refers to original design or development, variational design, conceptual re-design, or adaptive design (see *Design Process* (15)).

11. Design characteristic or feature

Any significant characteristic of the TS or MS that influences the constructional solution. *Design properties* (16) are a category of design characteristics.

12. Design documentation

Various written descriptions, graphical representations and other types of models of the partly or fully designed *machine system* (30), and *requirements* (17) placed on them in various forms and with various content. Roughly classified according to work flow during the *design process* (15). The following are listed in the usual sequence of generation during the design process, and comprise the main design documentation.

Design problem formulation

The *problem statement* (37) as an assigned specification, or list of *requirements* (17), as declared by the sponsor, who may be an external customer, the sales department, a company director, etc.

Design specification (17)

Clarified formulation of the problem, complete (as far as possible), orderly (sorted), quantified, and annotated with priorities. Includes all statements of requirements that are usable as *evaluation* (20) criteria. Can exist as a part of, or as an extension of, the contract documents.

Technical process (47) (TP) block diagram

Graphical representation of the *transformations* (49) of *operands* (33) through various operations (partial transformations) and their sequence as determined by the selected *technology* (51) (see Fig. 18).

Functional structure (27) schematic

Graphical representation of the partial functions required by the *technical process* (47), and their relationships. These exist in various forms, as block schematics (see Fig. 20), or hierarchical trees (see Fig. 21). They are intended to define the assigned *problem* (37) for each *machine system* (30) (MS) constrained within the *technological principle* (50), and are abstracted from other requirements.

Morphological matrix

Shows the appropriate *action principle*(s) (5) for partial *functions* (25), and the *organs* (34) capable of fulfilling each partial function (the *function-carriers* (26)), from which various combinations into a working whole can be selected (see Fig. 23). Each such combination forms a conceptual *principle* (35), on which concept sketches can be based.

Concept sketch

Represents in diagrammatic form, or written description, the rough *functional structure* (27) (and necessary *organs* (34)), or anatomical *structure* (43) built up from idealised components (see Fig. 24). *Form* (22), sizes and arrangements of the elements are not defined, even if they are in the limit very similar to those of the completed *machine system* (30).

Preliminary layout

Rough (first) iconic representation of the anatomical *structure* (43) (partly 'fleshed out' components) of a *machine system* (30), usually in the form of a free-hand sketch. The *features* (11) established at this stage are arrangements of elements, some details of their *forms* (22) and approximate dimensions, and class of materials.

Dimensional layout

Graphical representation of the anatomical *structure* (43) (components) of the *machine system* (30), that serves as starting point for *detailing* (18).

Drawing and parts list

Complete definition (description by verbal, numerical and graphical means, including computer-graphic data representations, micro-film, etc., as appropriate) of the *machine system* (30) by means of the *design properties* (16). Comprises detail drawings, sub-assembly and general assembly drawings, tooling and fixture drawings, cutaways, isometrics, lofting layouts, parts lists, operation and maintenance handbooks and manuals, etc.

13. Design manual (design catalogue)

Particular information source (or carrier) that should contain the necessary technological information in a comparative form (mostly as a table or matrix) suitable for a methodical approach to the designer's problems and other relevant aspects of design activity. VDI defines three types: object manuals, operations manuals and solutions manuals. Examples of solutions manuals are shown in Fig. 15, and consist of comparative listing of possible (functional) *organs* (34), e.g. methods of motion transmission, together with their equations, limitations, outline of anatomical *structure* (43), etc. Object manuals appear as data sheets or books, which are collections of experimental and theoretical results to assist design calculations, e.g. (53, 86)

14. Design methodology

General theory of the procedures for the solving of design *problems* (37). It involves both the general design strategy and also the tactical approach to individual portions of design work.

The main areas and forms of design methodology are:

Procedural model

Represents a concept for the overall execution of a *design process* (15). Idealised conditions are usually assumed for the *factors* (21) influencing the design process and the model is intended to be valid for all types of design *problem* (37).

Procedural plan

A clearly defined working procedure (plan), set up from time to time for a *problem* (37) (or class of problems) and at certain significant stages of the *design process* (15), taking into account the actual state of the preceding design operations and deviations from the idealised Procedural Model.

Procedural manner

An individual designer's (or design team's) ways and means of proceeding within a given Procedural Plan. It is strongly influenced by personal properties, characteristics and experience, e.g. *intuitive* (28) working.

Method

System of methodical rules that determine (classes of) possible procedures and actions which are intended to lead via a planned path to the accomplishment of a desired aim. Types may be classified according to method of thinking (*intuitive* (28) or discursive methods), or according to aim and application (methods of searching for solutions, methods of *evaluation* (20) or calculation, etc.).

Methodology

System of methods that may be used by an individual to attain a desired objective. For example, the way in which a teaching/learning process within an educational system is embodied, in the form of a curriculum or syllabus, and associated lecture outlines, case studies, problems, projects, experiments, demonstrations, etc.

Plan of action

Determines the activities (proposed tactics) for a concretely stated *problem* (37) situation with clear *constraints* (45) (boundary conditions).

Working method

The actual tactics and the typical activities of an individual designer in a particular problem situation and under given conditions.

Working principles

General instructions for appropriate behaviour in specific or typical problem situations, including methodical rules and general statements of objectives.

15. Design process—(DesP)

Consists of and describes the work flow of design, leading from the *problem* (37) formulation to the complete description of the *technical system* (48) (TS). Every design process may be structured into four main stages that can in turn be divided into a smaller or larger number of subsidiary activities, termed design steps. The main stages are:

- *elaborating* and clarifying the assigned problem formulation into a design specification: first stage of designing, in which one aims to develop a set of statements of the qualitative aspects of the problem as *requirements* (17) and desires, and where possible to quantify these.
- *conceptualisation*: main stage of design, in which a transition from requirements to *means* (31) takes place. The *functional structure* (27) is established as far as possible.
- *laying out*: quantitative design phase, in which the anatomical structure and most of the *design properties* (16) are established in a stepwise progression.
- *detailing* (18): last stage of design, intended to produce an exact description of the anatomical structure and all its parts. During the design process, the designer tends to use a *methodology* (14) to guide his actions and procedures. He will also refer to information-carriers such as *design manuals* (13) and other *specialist design information* (41). The extent to which this occurs depends in part on the *degree of originality* (10) of the desired solution.

16. Design property

Basic category of the *properties* (38) of the *machine system* (30), e.g. dimensions, material properties, etc. for each part as described during *detailing* (18), and which influence all other properties of the MS, e.g. running, operation, maintenance, etc.

17. Design specification

Clarified formulation of the *problem* (37), elaborated, complete (as far as possible), orderly (classified with respect to appropriate *aspects* (6), e.g. categorised as in Fig. 5), quantified set of *requirements* (17) for a machine system, annotated with priorities (fixed requirement, minimum requirement, desire, etc.). Includes all statements that can be used as *evaluation* (20) criteria.

18. Detailing

Last main operation of the *design process* (15), in which all *design properties* (16) are definitively established and the *technical system* (48) is fully described.

19. Effect

General descriptive term for the action of one object (its output effect) on another (its input). Effects may be direct or indirect. The result of an action is a reaction of the object subjected to that activity, in the form of a change of *state* (42) of some of its *properties* (38) following natural laws. In designing, certain effects are prescribed that are regarded as causes of the desired *transformations* (49), or the achievement of desired properties. From the viewpoint of the *machine system* (30), the output effect is its planned *purpose* (39)—its purpose *function* (25). The output effect performed by the machine system should not result in an appreciable change of state of the machine system itself, unless we are concerned with consumables (e.g. welding electrodes), or disposable goods (e.g. alkaline battery cells). The operational adjective *action* (1 to 5) is used to describe associated concepts.

20. Evaluation

Basic operation of assessing the quality of an object to be evaluated. This process consists of selecting evaluation criteria, determining appropriate *values* (52) for the *system* (45), and processing these to a combined value for the purpose of assisting a decision. Evaluation may be subjective or objective, emotional or intellectual, or a combination of these.

21. Factor (analogous to mathematical operator)

Influencing part of the process that exerts a direct *transformation* (49) effect, or affects the occurrences in some other way. A factor to a technical process may exist as a technical or machine system, the human being, or an external influence, e.g. from the environment.

22. Form

Fundamental characteristic of each *machine system* (30) that describes the geometry, proportions, etc., of the items. The forms of most machine systems are composed of combinations of basic geometric body shapes.

23. Form determination

A basic design activity in the search for the most suitable form (geometry, proportions, etc.) of the *machine system* (30) during the *layout* (12, 15) phase. This activity requires resolution of complicated interactions between *form* (22) and *function* (25), material, manufacturing methods, strength, human beings and cost (as probably the most important items). Knowledge about these contexts may be obtained from areas of 'design for function', 'design for materials', for 'manufacture', and for 'economics', that are derived in various engineering sciences. Guides to appropriate forms may be obtained by considering various *principles* (36) and *moderators* (24).

24. Form moderator

A *feature* (11) or quality that substantially influences the process of *form-determination* (23) in the *detail* (15) design phase, e.g. *action localities* (1), forces, materials, sizes, etc.

25. Function of the TS, MS

Different interpretations are in use. Two concepts are particularly useful in systematic *design methodology* (14):

(1) Function as the duty that the product must be capable of fulfilling, i.e. its *effects* (19) and actions, or the benefits or utility of the machine system (e.g. as in Value Analysis, (33, 57, 79)). Can also be stated as functional *purpose* (39), or aim.
(2) Function as a general connection between input and output (22) (analogous to a mathematical function) which is concerned more with the act of functioning, and method of function performance. The literature refers here to technical function (4, 5).

Because of this dichotomy, this term finds limited use within the context of this book, and then only in its first sense of capability of performance, not of its realisation.

26. Function-carrier

Means of realising a function. Every *organ* (34) is a function-carrier.

27. Functional structure

Description of the *structure* (44) of a *technical system* (48) (TS) or a *machine system* (30) (MS) with respect to the isolated *functions* (25) and their relationships. Degree of sub-division, *degree of abstraction* (7) of the description, and *representation* (12) either in block schematics (Fig. 20) or in hierarchical form (Fig. 21) can vary within wide limits. The hierarchical form (as a branched function tree) can generally not show the relationships between structural blocks located in different branches, it is therefore sometimes termed a decomposition (sub-division) of function (9).

28. Intuition—intuitive method

Finding solutions on the basis of 'ideas' that emerge suddenly and in surprising completeness. We are mainly dealing with experiential thought, in which previously collected information and experience are worked into a synthesis in a sub-conscious process that should not be explained as irrational. The opposite is *discursive* thought.

29. Machine

A *system* (45) based generally on mechanical *principles* (35) that is capable of exerting a somewhat complex *transformation* (49) and propulsion *effect* (19). Various overlapping categories are generally recognised: process machines, energy conversion machines (e.g. motors and other prime movers), signal processing machines, etc. A machine comprises assemblies and parts. Previous characterisations that are no longer fully valid include movement, and capability of performing useful work by conversion or transmission of forces.

30. Machine system—(MS)

A category of *technical systems* (48) (TS). MS is used as a collective term for machines, apparatus, devices, transportation media, etc. These are mainly products of mechanical engineering manufacture (all others are more appropriately classed as Technical Systems). Inputs to machine systems are materials (e.g. fuels, lubricants, etc.), energy and information signals, that are transformed into *effects* (19) (output actions), usually by mechanical means. These actions serve to perform *transformations* (49) in the *technical process* (47) (see Fig. 1).

31. Means

General designation for the solutions to a *problem* (37), particularly an object, apparatus, or mechanism of physical/technical nature, but can also refer to mathematical, financial, etc., including software and hardware systems for information filing and retrieval, modelling, representation, reproduction, calculation, etc.

32. Mode of action

Nature of those interactions of *organs* (34) that are intended to achieve an *effect* (19), and that are based on natural phenomena. In most cases an 'if, then' (cause, and consequent effect) description of the mode of action is used. This is a typical form of description of *machine systems* (30) as is used in educational textbooks.

33. Operand

An object that undergoes a change of *state* (42) or other *transformation* (49) within a *technical process* (47) as a result of *effects* (19) exerted upon it by one or more *technical systems* (48). Such objects may include biological (including human) and material objects, quantities of matter, energy, or information signals.

34. Organ

A component of the *machine system* (30) that forms a *function-carrier* (26) and exerts a certain *effect* (19) or performs a *function* (25). An organ is not necessarily identical to a physical sub-assembly (e.g. a multi-plate clutch), but may be composed of connective parts of two or more sub-assemblies (e.g. the splined holes in those clutch plates, together with the splines on the shaft). We differentiate: transformation, propulsion, auxiliary, control, and also connection and support organs.

35. Principle

The source, or basis, or law (e.g. of nature), or primary element, or fundamental truth (e.g. an idea), from which the other laws, elements, etc., may be derived or on which they are dependent, may also refer to idealised methodical rules to guide the execution of the *design process* (15), or the guiding thought (e.g. a guiding principle for representation, such as 'third-angle projection').

36. Principle of form determination

A basic concept from which a particular *form* (22) can be developed, e.g. 'aerodynamic shape' as a *principle* (35) for *form-determination* (23) of motor vehicle bodies.

37. Problem (problem statement)

An assignment for which a solution is to be found. In a general *design process* (15), the problem statement is usually given in a form as it is perceived and formulated by a *sponsor* (12). In smaller portions of design work, the design engineer may formulate a problem for himself.

38. Property

Any attribute or characteristic of a *technical system* (48): performance, *form* (22), size, colour, stability, life, manufacturability, transportability, suitability for storage, *structure* (43), etc. Every technical system (TS) is deemed to be a carrier of all properties, these properties exist in that form only because the TS exists. Their totality represents the *value* (52) or quality of the system. Properties may be important or unimportant, variant or invariant in time, external or internal.

The basic category of properties in a Technical System (TS) is that of the fundamental *design properties* (16), namely the anatomical *structure* (44) (component parts and their arrangement), and also form (geometry), dimensions, material and manufacturing methods, tolerances and surface properties of the separate components. These primary properties determine all other properties of the system such as performance, life, etc. The quality of each property is expressed in its value, its nature (e.g. colour), state of embodiment (e.g. sky blue), measure (e.g. colour number 6-16P from a manufacturer's standard chart), as unique to that system. Important properties, mostly those connected with the purpose of the machine system, are termed *parameters*.

39. Purpose

The purpose of *technology* (51) is generally the attainment of certain desirable *states* (42) of things, i.e. of *operands* (33), that should serve to satisfy certain human needs at a particular time, and under given conditions, including the existing state of the art at that geographic locality. The desired state of an operand is achieved by applying suitable technical *means* (31).

40. Relative value

A quantitative statement of quality of the *system* (45), relative to a given technical or economic ideal, e.g. where an absolute measurement scale does not exist, or is inconvenient.

41. Specialist design information

Every item of knowledge of technical, economic or organisational nature, that can be of use during the *design process* (15), including instructions about procedures, etc.

42. State

The state of a *system* (45) is the totality (analogous to a mathematical vector) of all *values* (52) of its *properties* (38) at a given *instant in time*.

43. Structure

General term for a perceived orderly arrangement and the ordering relationships between elements in a *system* (45) from certain viewpoints, and its description or definition. In particular cases, topology and microstructure can be used in this description. Examples: (1) a sentence in language has a structure, described by syntax, grammar, parsing, etc. (2) a mechanism has a physical (anatomical) structure with changing geometry, described by its parts, assembly instructions, degrees of freedom, etc. (3) a framework's structure is static, leading to an interchangeable usage of the word 'structure' for the description and for the physical entity, e.g. as in 'structural engineering'.

44. Structure of machine systems

This may be considered from the two viewpoints of *function* (25), or of construction (manufacture).

The functional viewpoint is concerned with actions and behaviour of the system elements, the *organs* (34), and is generally dynamic. It refers to a *functional structure* (27) that describes the MS by means of its functions, or of the organs that perform those functions. Hierarchy of *complication* (8): organ, organ device, organ system. In each machine system various typical classes may be differentiated, namely transformation- (main-, or work-), auxiliary-, propulsion (drive-), control-, connection- and support-organs. Relationships are of a functional nature (input, output, coupling). The functional structure may be represented in various levels of *abstraction* (7), e.g. as organ structural schematics (see *Design Documentation* (12)).

From the constructional viewpoint one can define an *anatomical structure* (43), more familiar to manufacturing personnel, that describes the *machine system* (30) by means of its constructional elements. Hierarchy of *complication* (8): constructional element, detail part, machine element, sub-assembly, assembly, machine, equipment or plant. This largely static viewpoint, similar to a biological anatomy, is concerned particularly with the demands of realisation, i.e. manufacture, and with the *intricacy* (9) of the components.

Relationships between functional organs and constructional elements may vary: they can be identical, or an organ can consist of a number of constructional elements, or a constructional element may contain parts of a number of organs. For instance, a lathe bed as a machine component contains a portion of the longwise motion organ (the slideways), and of the lubrication organ (the oil channels and holes), and is at the same time the main connective and support organ.

45. System

A set of elements and their relationships within a clearly defined *boundary*.

Various totalities may be regarded as a system, e.g. a *process* (47) as a set of operations, a device as a set of its components, or of its *properties* (38) or *functions* (25) (see *Functional Structure* (27)). For example, an i.c. engine connecting rod may be regarded as a system, with its boundary drawn around its outer surface and the big- and little-end bearing surfaces. It would be equally valid within the context of a specific design problem to consider a complete gas turbine engine, or a single aircraft, or an international air traffic network, as a system.

Every system has the characteristics: input, output, performance, system behaviour, system structure, system boundary, and environment of the system.

Partial system

A *partial system* (Sub-System) is a relative term for a component part of a system, at various levels of *complication* (8).

System boundary

A *system boundary* is an arbitrary and imaginary dividing line drawn between the system under consideration and its environment. Connections between the system and the environment are provided by receptors (to accept the inputs) and effectors (to deliver the outputs), and these are components of the system.

Environment of the system

The *environment of the system* consists of all elements not within the system boundary. It is necessary to discriminate between the local (relevant) environment that has a relationship to the system, and the remote (not relevant) environment with no relationship to the system. In the analysis of a system, the local environment must be considered, one can denote this as the *space* into which the system is placed. Time must also be considered.

46. Technical artefact (see Technical System (48))

47. Technical process (TP)

An artificial process or procedure in which the *states* (42) of material objects, biological objects, energy, and information (the *operands* (33) of the TS) are *transformed* (49) in a planned and goal-oriented way under the influence of human beings and by the *effects* (19) exerted by technical *means* (31) (see *Technical System* (48)). The obtained states of operands should (directly or indirectly) serve the satisfaction of human needs (*Purpose* (39)).

The necessary operations (or partial transformations) and their sequence is established from the selected *technology* (51), which is based on natural laws and phenomena. A system *boundary* (45) divides this process from its environment, and this is characterised in locality (space) and time.

48. Technical system (TS)

A general category of artificial deterministic *systems* (45) that perform the necessary *effects* (19) to achieve the *transformation* (49) of *operands* (33). It is a collective term for all *machine systems* (30), devices, apparatus, equipment, plant, etc., from any branch of engineering (e.g. constructional systems, electro-hydraulic systems, control systems, communications systems, etc.).

49. Technical transformation (change, conversion)

The process of performing a change of *state* (42) of the *operands* (33) towards a defined target state, in *technology* (51) mostly from an existing undesired state into a desired state.

50. Technological principle

A recognised and established *principle* (35) from which a specific technology may be derived for design purposes. The technological principle can in the limit be identical with the *action principle* (5), if the appropriate *technical process* (47) describes the *purpose* (39) of the *technical system* (48), as for instance for energy conversion machines.

51. Technology

Ways and *means* (31) of exerting *effects* (19) on an *operand* (33) to bring it into a desired *state* (42). The basis for a search for appropriate effects is an extensive knowledge of natural phenomena and the associated sciences, particularly of manufacturing technology, materials science, metallurgy, production technology. A technology is in practice defined by the necessary operations and their sequence.

52. Value (also quality, measure, composition)

The *means* (31), *purpose* (39), performance and assessment of a *system* (45) refers either to a single *property* (38) (e.g. the value of acceleration, size, packing possibility) or to more complex values, e.g. total value, technical value, usage value or benefit. The assessment of value can use qualitative or quantitative information to produce a valuation statement (see also *relative value* (40)). One of the purposes of natural sciences is the conversion of qualitative to quantitative information and models (quantification). For instance, colour is a subjectively perceived property that may be described in terms of vibrational frequencies. A different solution is the determination of accurate measures (scales), such as the standard colour definitions (e.g. to DIN 6164, BS 4800, or other appropriate specification), which prescribes a system of standardised colour tones. A desired colour is described in this system by a three-digit colour characteristic number (various term combinations are in use, e.g. chroma, saturation, intensity; hue, saturation, darkness; hue, greyness, weight; etc.).

Index

Note: bold type indicates main reference in text and glossary
italic type indicates subsidiary reference in summaries, tables and figures

Abstract (-ing, -ion), 2, 5, 9, 22, *22*, 24, 29, 34, *40*, 41, *43*, 45, 50, 51, 53, 56, 57, 65, 76, 95, 96, 97, **99**, 101
Action, 1, 15, 96, **98**, 99
  chain, 15, *26*, 54
  locality, 12, 13, **17**, 19, 53, 54, 55, 58, 59, 65, **98**
  principle, 13, 15, 18
Activity, 27, 32, 38, *42*, 48, 63, 102, 103
Adaptive, *38*, 45, 100
Adjustment, *25*
Aesthetic, 64, 75, 96 (see also Appearance)
Aggregation, 38, *38*, 65
Agreement, 4, 65
Alternative, 19, 51, 53, 55, 58, 59, 68
Analogy (-ue, -ous), 18, 22, 30, 41, 65, 76
Analy-se (-sis), 1, 28, 49, 111
Anatomical structure, 14, **17**, 18, *25*, *26*, 34, 36, *43*, 45, 50, 53, 54, 56, 57, 58, 59, 60, 61, 65, 101, 102, 104, 108
Apparatus, 45, 111
Appearance, 15, *16*, *41*, 96 (see also Aesthetic)
Application, *38*, 44, 52, 53
Arrangement, 15, *16*, 18, 22, 24, 26, 55, 56, 57, 58, 50, 96, 101, 108, 109
Artefact, 40, 41, 97, **111** (see also Device; Mechanism)
Artistic, 29
Aspect, 29, 34 (see also Classifying aspect)
Assembly, 17, 18, *25*, *41*, 59, 96, 97, 99, 100, 107, 109, 110
  drawing, 17, 102
Assigned, 2, 30, 32, *42*, *43*, 89, 96, 100, 101, 104, 108 (see also Problem)
Association, 28, 39
Assumption, 36
Attribute, *38*, 64, 108
Automation, 6, *11*, 18, 21, *26*, 51, 52
Auxiliary, 7, *11*, 14, 18, *25*, *26*, 52, 53, 108, 110

Basic, 32, *33*, 48, **62**, 65, 68 (see also Operation)
  shape, 58
Batch, 45
Behaviour, 17, *25*, 37, 40, *43*, 55, 68, 95, 110
Black box, 18, 50, 51, *82*
Block, 48, 50, 57, *84* **101**, 106 (see also Diagram; Schematic)
Bought-out, 19, 58 (see also Component)
Boundary, 7, 17, 36, 50, 51, 52, 53, 103, 110, 111
Brainstorming, *38*, 39, 64, 65
Branch, 24, *26*, 30, 32, 34, 111

CAD, 42, 45, 61
Calculation, 58, 60, 61, 102, 107
Capability, 12, 14
Carrier, 15, *25*, 34, 77 (see also Function)
Case study, 47, 49, 50, 54, 58, 60, 61, **77**, 89
Catalogue, 59, 73 (see also Design)
Category, 64, 100 (see also Class)
Cause (-ality), 1, 5, 15, 16, *25*, 63, 105, 107
Chain, 15, 25 (see also Causality)
Change, 5, *10*, *11*, 21, 111 (see also Transform)
Characteristics, 15, 17, 21, 22, 24, 36, 48, 49, 69, 96, 99, 100, 105, 108, 110
Check (-ing), 12, 33, 45, **62**, **74**, 75, 97
Check-list, 40, 49
Class, 8, *11*, 14, *16*, 22, 23, *25*, 50, 55, 56, 96
Classification, 7, 22, *22*, 23, 49, 50, 63, 73
Classifying aspect, **24**, *25*, **99**
Coding, 41
Combin-ing (-ation), 16, *25*, *39*, *41*, 52, 56, 69, 72, 101
Command, 12, 18
Communicat-ing (-ion), 33, 40, 41, 62, 76, 96
Compatibility, 56

113

Completeness, 14, *35*, 36, 52, 64
Complication, 17, 18, 22, *26*, 36, 45, 53, 54, 68, 74, *99*
Component, 8, 17, 18, *26*, 30, 32, 95, *96*, 98, 99, 100, 101
Composite, 16 (see also Value)
Computer (CAD), 42
Concept, 2, 27, 28, 32, 36, 45, 53, *54*, 56, 57, 58, 76, **97**, 102
 sketch, 54, *85*, **101**
Conceptual, 45, 54, 101 (see also Design)
Conceptualise (-ation), 32, 48, 54, 56, 75, **104**
Conclusion, 7, *11*, 14 (see also Phase)
Concrete, 14, 22, 24, 36, 55, 57, 65, **96**, 97, 99
Concretise (-ation), 14, *22*, 24, 34, 36, *43*, 44, 45, 56, 58, 60, 61, 62, 68, *97*
Condition, 7, 28, 36, 45, 49, 53, 58, 65, 102, 103
Conformity, 36, 64, 69
Connect-ion (-ivity), 7, *11*, 14, 18, *25*, *26*, 32, 52, 106, 108, 110, 111
Conscious, 28
Constructional, 14, 15, 56, 57, 58, 60, 61, **96**, 97, 110 (see also Element)
Contract, 49, 50, 69, 101
Control, 6, 7, *11*, 12, 14, 18, *25*, *26*, 29, 30, *40*, *43*, 45, 48, 52, **97**, 108, 110 (see also Feedback; Regulation)
Controllable, 28
Convergence, 34
Cost, *41*, 59, 61, 105
Cost-benefit, 61, 71
Coverage, 68
Create (-ivity), 3, 28, 31, 65
Criteria, 16, *25*, 36, 51, 53, 68, 69, 105 (see also Evaluation)
Critical path, 37, *39*
Critic-ism (-al), 34, *40*, 49, 54
Customer, 31, 32, 49, 50, 64 (see also Sponsor)
Cycle (-ic), 32, *43*, 45

Data, 29, 73, 74
Deadline, 45, 74
Decide (-sion), 3, 6, 16, 28, 33, 36, 51, 61, 62, **68**, *70*, 97, 105
Decompose (-ition), 30, 45, 53, 54
Deficiency, 63, 65
Degree of freedom, 96, 109
Descartes, *38*
Descri-be (-ption), 27, 31, 41, *42*, 48, 61, 97, 100, 101, 102, 103, 104, 105, 107, 109
Design
 catalogue, 56, *66*, *67*, 68, **102**
 characteristic, 15, *26*, *42*, 49, 51, 55, 57, 60, 61, 65, 80, **100**
 engineer, 3, 15, 19, 24, *26*, 31, 32, 37, 40, *43*, 44, 45, 64, 75, 76, 108
 engineering, 2, 27, 29, 62, 64, 71, 72, 89, 97, 112
 manual see Design catalogue

Design (*cont.*)
 process (DesP), 14, 17, **27**, 29, 30, 31, *31*, 32, 33, *33*, 37, 41, *42*, *46*, *47*, 74, 89, **103**
 property, 15, *16*, 21, *25*, *26*, 34, *42*, 57, 59, 71, 76, **104**
 specification, 20, 33, 49, 50, 63, 65, **101**, **104** (see also List of requirements)
Designer see Design engineer
Detail (-ing), 45, 48, 53, 56, 58, 60, **61**, 62, 75, 96, 102, **104**
Determine, 53, 97
Deterministic, 15, *25*, 111
Development, 6, **21**, *26*, 27, 45, 64, 89, 100
Device, 3, 95, 110, 111
Diagram, 80, 101
Difficulty, 36, **100**
Dimension (-ing), *16*, 19, 32, *38*, 58, 59, 60, 101, 104, 108
Dimensional layout, 58, **59**, 61, *88*, **102**
Discursive, 39, 44, 50, 65, 103
Distribution, 6, *11*, *16*, 52 (see also Involvement)
Disturbance, 7, *11*, **16**, *25*, 54 (see also Secondary)
Division, 39, *41*, 65
Document (-ation), 3, 22, 41, *41*, *42*, 73, 76, **100**
Doubt, *39*, 64, 74
Drawing, 41, 61, **102** (see also Detail)
Duration, *43* (see also Time)

Economy (-ics), *16*, *39*, *40*, 41, *43*, 49, 61, 64, 69, 71, 109
Effect, 2, 5, 6, 8, *10*, 12, 13, 14, 15, *16*, 17, 18, 19, *24*, *25*, 31, 34, *43*, 45, 50, 52, 53, 54, 55, 63, 65, **96**, **104**, 105
Effective, 3, *40*
Effector, 12, 19, *26*, 54, 97, 98, 111
Efficiency, 44, 69
Elaboration, 33, **49**, 52, **62**, 63, 64, **104**
Element, 14, *16*, 42, 97, 109, 110 (see also Constructional)
Elementary design property, *16*, 34, 36, *42*, *43*, 45
Embodiment, 15, 55, 71, 97, 103 (see also State)
Emission, 17, *25* (see also Waste)
Energy, 8, *10*, *11*, 12, 14, 18, 19, *25*, 52, 95, 96, 107, 111, 112
Environment, 5, 7, *11*, 15, 16, 17, *26*, 32, *43*, 54, 95, 110, 111
Equipment, 12, *25*, 45, 99, 110, 111
Ergonomic, 15, *16*, 60, 75
Error, 28, 29, 39, 74, 75, 97
Establish, *10*, 14, **17**, 50, 51, 52, 53, 54, 56, 57, 58, 59, 61, **97**
Ethical, 5
Evaluate (-ion), 3, 16, *25*, 28, 29, 33, 36, *38*, 39, 45, 51, 52, 53, 56, 57, 59, 60, 61, *68*, *70*, 74, **105**
Evaluation criteria, 20, 33, 49, 50, 57

Index

Example, 49 (see also Case study)
Execution, 7, *11*, 14, 29, 52
Experience, 1, 54, 58, 64
Experiment, *38*, *39*, 60, 71, 75
Expert, 68, 73, 75 (see also Syndicate)

Factor, 7, 8, *11*, 17, 20, *26*, 30, 36, 37, 41, 75, **105**
Failure, 30, *41*, 58, 62, 63
Family, *11*, 44, 54, 96, 99
Feasibility, 49, 67, **68**
Feature, 24, 36, 48, 49, 96, **100** (see also Design characteristic)
Feedback, 97
Fidelity, 36
Filing, 73, 74, 107
Finance, 64
Fixation, 28, 56, 68 (see also Prejudice)
Fixed, 50, 63, 104 (see also Requirement)
Fixing, 41, 65 (see also Characteristics)
Fixture, 45, 102
Flow Diagram, 20, 48
Form, 2, 8, *11*, 12, 15, *16*, 21, 24, *25*, 44, 51, 57, 58, 59, 60, **96**, **105**
Form-design, Form-determination, Form-giving, 32, 58, 59, 60, 75, **105**, **108**
Formulation, 28, *40*, **100**, 101, 103, 108
Function, 6, 14, *16*, 17, 41, *41*, *43*, 45, 50, 51, 53, **96**, 97, 98, **106**
Function tree, 54
Function-carrier, 17, 18, 54, 55, 56, 96, **106** (see also Organ)
Functional
 structure, **14**, 18, 19, *25*, 36, 50, 51, 52, 53, 54, 57, **101**, **106**
 unit, 53 (see also Assembly)
Functioning, 14, 106

General procedural model, 44
Generat-ing (-ion), 20, 21, *26*, *42*, 97, 100
Genus, *11*, 24, 57, 96, 99
Glossary, 4, 47, **95**
Group (-ing), 15, 51, 52, 53, 100

Heuristic, **28**, 30, *43*, 45
Hierarchy (-ical), 9, 18, 22, *25*, 32, 52, *84*, **96**, 101, 106, 110
Human, 8, *10*, 41, 52, 71, **97**, 111
Hybrid, 24

Iconic, 57, 76, 101
Ideal, 16, 17, 36, 37, 71, 109
Imagination, 54, 97
Implement (-ation), 1, 96, **97**
Improve (-ment), 21, 34, 52, 53, 57, 59, 71
Incubation, *38*, *41*, 47

Information, 8, *10*, *11*, 12, 14, 18, 19, *25*, 30, 31, 33, 37, 40, 41, *42*, *43*, 44, 45, 49, 50, 52, 55, 58, 62, 63, **72**, **73**, 76, 95, 96, 102, 104, 107, 109, 110 (see also Specialist design information)
Innovation, 21, *26*
Input, 7, 8, *11*, 12, 13, 14, 17, 19, 21, *25*, *26*, 34, *43*, 54, 95, 96, 98, 104, 106, 107, 110, 111
Inspect (-ion), 3, 7, 29
Instruction, 38
Instrumentation, 6
Interact (-ion), 16, 17, 105, 107 (see also Connection; Relationship)
Intricacy, 32, **100**
Intuit-ive (-ion), 3, 28, 29, 44, 48, 50, 65, **106**
Invention, *38*, 68
Involvement, *26*, 51, 100 (see also Distribution; Human)
Irrational, 29, *42*, 106
Iterat-ive (-ion), 34, 36, *38*, *43*, 65

Jig, 45
Judgement, 16, 61 (see also Evaluation)

Key-word, 63, 77
Knowledge, 2, 7, *10*, 15, *25*, *26*, *43*, 54, 63, 72, 73, 89

Laws of nature, 5, 55, 64, 72, 74, 105, 111
Layout, Laying out, 45, 48, 57, 58, 59, 60, 76, **104**
Learn, 2, 29, 77
Life, 20, *20*, *21*
List of requirements, 20, 36, 49, 63, 65, 68, *83*, 100 (see also Design specification)
Loading, 58
Locality see Action locality
Location, 7, 8, *11*, 12

Machine, 12, 18, *25*, 45, 99, **106**, 110
 element, **18**, *19*, 26, 96, 110
 system (MS), 17, 18, *24*, *25*, 49, 57, 64, 65, 71, 76, 89, **107**, **110**
Magnitude, *25* (see also Value)
Main, 7, 51, 52 (see also Transformation)
Management, *11*, 21, 32, 37, 40, *43*, 45, 64, 72, 97
Manufacture, 2, 7, 8, 15, *16*, *25*, *41*, 45, 57, 59, 60, 61, 62, 64, 97, 100, 105, 108, 110, 112
Mapping, 41
Market, 37, *39*, *41*, 49, 64
Material, 7, 8, *11*, 12, 14, 15, *16*, 19, *25*, 52, 55, 59, 60, 95, 96, 101, 107, 108, 112
Means, 1, 2, 6, 8, *10*, 18, 34, *43*, 63, 64, 65, 106, **107**

Index

Mechanisation, 6, *11*, 21, *26*, 51, 52
Mechanism, 12, **96**, 97, 98
Mental, 2, 27, 72, 97 (see also Activity; Process; Thinking)
Method, 1, 3, 28, 30, **37**, 38, *38*, *43*, 65, 74, 89, **103**
Method '6-3-5', *39*, 68
Methodical, 22, *40*, 43, 44, 48, 65, 76, 103, 108
Methodology, 22, 29, **102**, **103**
Minimum, 50, 63, 104 (see also Requirement)
Mode of action, 13, *14*, 15, *16*, 17, 21, 24, *25*, *26*, 53, 54, 55, 56, 89, **107**
Model (-ling), 6, 8, *11*, 13, 20, 22, 24, 25, 27, 31, 37, *39*, *43*, *46*, 51, 76, 89, 97, 100, 107
Modification, 30, 34
Morphology, *39*, 54, 55, *85*, **101**
Motivation, 28, 69
Mutation, 36, *43*, 44 (see also Variant)

Nature (-al), 5, *25*, 55, 99, 112 (see also Laws of Nature)
Need, 2, 5, *10*, 63, 97, 109, 111
Negative, 50 (see also Requirement)
Network, *39*, 64
Neutral, *43*, 44, 49 (see also Object-independent)
Novel, 56, 100

Object, 8, *10*, 45, 97, 102, 107, 111
Object-independent, 29, 30, 44 (see also Neutral)
Objective (-ity), 1, 16, 28, 63, 68, 69, 71, 105
Obligatory, 14 (see also Fixed)
Observe, 5
Operand, **8**, *10*, *11*, 12, 19, *24*, 45, 51, 65, **96**, 97, **107**
Operation, 3, 7, 8, *10*, *11*, 15, *16*, 20, *26*, 28, 30, 37, *41*, *42*, *43*, 51, 52, 101, 102, 103, 104, 110, 111, 112
Operator, 12 (see also Human)
Optim-um (-isation), *25*, 27, 36, *40*, *43*, 44, 51, 52, 53, 59, 61, 65, 68, 69
Order, *40*, 50
Organ, **17**, 18, 19, *24*, *26*, 54, 55, **96**, 97, **98**, **107**, 110
Organ structure, *26*, 36
Organising (-ation), 7, *11*, *16*, 29, 50, 64, **95**, **96**, 97, 109
Orientation, *40*, 48, 58
Originality, 36, 68, 74, **100**
Origination, 20, *20*, *21*, *26*, 89
Outline, 19 (see also Drawing)
Output, 7, 8, *10*, *11*, 12, 14, 18, *25*, 34, 50, **95**, **96**, 98, 104, 106, 110, 111

Parameter, 7, 15, 16, 64, 65, 68

Part, 17, 18, 25, **102**, 109, 110 (see also Component)
Partial, 7, *11*, 55, 58, 76, 101, 110 (see also Function; Process; System)
Performance, 71, 108, 110, 112
Personal working mode, **39**, *43*
Personality, 64
Phase, 20, 29, 36, 37, 45, 48, 52, 62
Phenomenon, 5, 15, *25*, 54, 55, 99, 112
Phylum, *11*, *22*, *23*, 96
Plan (-ing), 1, 6, *10*, 16, *16*, 29, 30, 37, 38, 40, 45, 50, 64, **97**, 102, **103**
Plant, 12, *25*, 45, 52, 65, 99, 110, 111
Policy, 21
Pre-conscious, 28 (see also Conscious; Rational)
Prejudice, 28, 56, 68 (see also Fixation)
Preliminary layout, 36, *57*, 58, 59, 60, *88*, **101**
Prepare (-ation), 7, *11*, 14, *16*, 33, 48, 50, 52, 62, 64, *72*, 73
Principle, 13, 15, 18, 44, 55, 56, 63, 75, **99**, **108**
Priority, 28, 50, 104
Probability, 63, 74
Problem, 2, 3, 28, 32, 37, *40*, 48, 49, 62, 63, 64, 68, **100**, 102, **108** (see Assigned)
  axis, 34
  solving, 27, 32, 89
Procedural manner, **37**, **103**
  model, 32, 34, 36, **37**, *43*, 44, 48, **102**
  plan, **37**, *43*, **102**
Procedure, 3, 6, *11*, 29, 37, 38, *40*, *46*, *47*, 68, 74, 102, 103, 109, 111
Process (-ing), 2, 5, 6, 8, *11*, 20, *26*, 31, 51, 68, 72, 75, 89, **95**, **96**, 97, 111
Product, 62
Production, 2, 28, 37, 45 (see also Manufacture)
Professionalism, *41* (see also Ethical)
Profitability, 69
Progress, *43*, 64, 65, 99
Project, 64, 65, 75 (see also Teamwork)
Property, 8, 15, *16*, 17, 21, 24, *25*, 34, 36, *38*, 39, 62, 64, 65, 68, 72, **96**, 97, 105, **108**
Prototype, 45, 75, 76
Provide, 33, 62, **72** (see also Information)
Psychology, 30, *43*, 45
Purpose, 12, 18, 19, *43*, 50, 51, **109**

Qualitative, 6, 109, 112
Quality, 15, 16, *16*, *25*, 36, *41*, 43, 61, 64, 69, 71, 72, 105, 108, 109, 112
  control, 29
Quantify (-ication), 1, 16, 50, 63, 71, 104, 112
Quantitative, 6, 109, 112
Questioning, *38*, 39, 64
Questionnaire, 49

Rating, 68, 69 (see also Weighted)
Rational (-isation), 3, 27, 28, 29, *42*, 45, 48, 49, 63

Index

Realise (-ation), 1, 5, 6, 7, *25*, 36, *41*, *43*, 52, 63, 64, 71, **97**, 98, 100, 105, 110
Receptor, 19, *26*, 97, 98, 111 (see also Effector)
Record (-ing), 3, 7, *40*, 48, 55, 60-1, 68, 73, 76
Recycling, 17, *25*
Regulation, 37, *41*, 97 (see also Standard)
Relationship, 14, *16*, 34, *43*, 44, 55, 56, 57, 76, 101, 106, 109, 110, 111
Relative, 18, **109**
Relative-strength diagram, 71
Release, 45, 61, 76
Reliability, 16, *16*, 58, 72, 99
Repeat part, 58 (see also Re-use)
Report (-ing), 7, 61
Representing (-ation), *11*, 22, 24, 31, 32, 33, 40, 41, *41*, *43*, 48, 50, 52, 56, 57, 59, 60, 62, **76**, 89, 100, 101, 102, 107, 108
Requirement, 5, *10*, 15, 31, 34, 36, 40, *42*, *43*, 44, 49, 50, 51, 57, 58, 63, 64, 65, 69, **97**, 101 (see also Design specification; List of requirements)
Research, 29, **30**, 32, 34, *39*, 72
Resource, 1
Restriction, 50
Retrieval, 73, 74, 107 (see also Filing)
Re-use, 58
Review, 3, 7, 45, 47, 49, 53, 56, 61, 68, 75

Scale, 57, 59, 60, **69**, 96, 112
Schematic, 27, 52, 57, 76, **101** (see also Block; Diagram)
Scien-ce (-tific), 22, *25*, 29, 112
Scientific scepticism, *39*, 64, 74
Search, 3, 33, *38*, *40*, *42*, *43*, 45, 51, 62, **65**, 68, 73, 103, 112 (see also Solution)
Secondary, 8, 17, *25* (see also Disturbance; Effect; Input; Output)
Sensitivity, 16, 68
Sequence, 3, *10*, 12, 22, *26*, 28, 30, 37, 38, 51, 52, 62, 111, 112
Significance, 68
Sketch, 2, 57, 59, 101 (see also Concept)
Slip clutch, 55, 56
Society, 37
Solution, 1, 19, 39, 50, 59, 62, 63, 64, 99, 100, 102, 107, 108
Space, 7, 12, *16*, *16*, 54, 111 (see also Environment; Location)
Specialist, 61 (see also Expert)
Specialist design information, 31, 59, 72, **109** (see also Information)
Sponsor, 49, 108 (see also Customer)
Stage, 3, 14, 48, 63, 104 (see also Phase; Step)
Standard, *16*, 37, 58 (see also Regulation)
State, *10*, *43*, 45, 48, 49, 63, 73, 75, 89, 96, 99, 102, 103, 105, **109**
  of the Art, *11*, 21, *26*, 49, 63, 109
  of assembly, 15
  of embodiment, **15**, 24, *25*, 108

Step, 3, 30, *39*, *43*, 44, 48, 62, 104 (see also Phase; Stage)
Storage, 8, *16*, 41
Strategy, 34, *43*, 74, 89, 102
Strength, *16*, 41, 58, 60, 105 (see also Calculation)
Stressing, 60, 75
Structure, 7, 8, *10*, *11*, 12, *16*, 17, 31, *43*, 44, 51, 52, 65, 89, **95**, 96, 97, 104, **109**, **110** (see also Anatomical structure; Functional structure; Organ structure)
Study, 67
Sub-divide (-sion), 7, 18, 54
Subjective, 16, 68, 105
Substantiation, 44, 60 (see also Verification)
Supervise, 29, 97
Support, 7, *11*, 14, 18, *25*, 108, 110
Surface, *16*, 59, 60
Symbol (-ic), 57, 76
Syndicate, 61, 75
Synectics, 39, *39*, 68
Synthesis, **56**
System, 8, 17, 19, 31, *39*, *40*, 41, 64, 73, **95**, 96, **110**, 111
Systematic design, 27, 47 (see also Methodical)
Systems thinking, 39

Tactics (-al), 30, 37, *38*, 40, *40*, *43*, 89, 102, 103
Task, 12 (see also Purpose)
Teamwork, 3
Technical process (TP), **5**, *10*, 12, 18, 22, *22*, *24*, *26*, *42*, 45, 51, 52, 53, 54, 89, **101**, **111**
Technical system (TS), **12**, 20, 21, *24*, *25*, 27, 30, 31, 34, 36, *42*, *43*, 45, 48, 52, 53, 54, 61, 62, 63, 65, 71, 72, 111
Technological principle, *10*, 13, 24, 50, 51, **112**
Technology, 2, 5, **6**, 8, *10*, 12, 13, 21, 22, 52, 54, 100, **112**
Term, (-inology), 3, 22, 32, 47, 72
Testing, 7, 45, 60, 76 (see also Experiment)
Theorise, 2
Thinking, 27, 28, *33*, 39 (see also Mental; Activity)
Throughput, 8, 12, 89, 96
Time, 7, 8, *11*, 12, 16, 54, 111
Tolerance, 15, *16*, *41*, 50, 59, 60, 63, 108
Tool (-ing), 6, 7, *41*, 45, 102
Transform (-ation), 2, 6, 7, 8, *10*, 12, 14, 18, *24*, 31, 51, 89, 95, 96, 97, 98, 107, 108, 110, **111** (see also Change)
Transmission, 58
Transport, 8, *16*, *41*, 75
Trend, 21, 64
Type, *22*, *23*, 24, 57, 96, 99

Uncertainty, 36
Unconscious, 28 (see also Conscious)
Undesired, *25* (see also Disturbance)

Value, 5, 15, 16, 21, 25, 29, 53, 69, 71, 72, 96, 97, **109**, **112**
   analysis, 37, *39*, 58, 60, 61, 76, 106
   engineering see Value analysis
Variability, 15, 50, 57
Variant, 19, 21, *26*, 28, 41, 56, 59, 100 (see also Mutation)
Vector, 69
Verify (-ication), 33, 36, 45, 52, 57, 59, 60, 62, **74**, 97
Visualising, 41, *42*, 48, 76

Waste, 17 (see also Emission)

Weak point, 53, 57, 59
Weighted, 68, 71 (see also Rating)
Welding positioner, 57, **81**
Work
   organ, 18, *26*
   study, 28, 45
Working
   conditions, 28, 32, 34, 37
   means, 31, 37, **41**, 43, 45, 58, 89
   methods, 29, 31, **38**, 43, 64, **103**
   principles, 37, 38, **40**, 40, 43, 50, **103**
   situation, 40
Workshop, 61, 62